Ultra Wideband Systems

Ultra Wideband Systems
Technologies and Applications

Roberto Aiello
Staccato Communications

Anuj Batra
Texas Instruments

AMSTERDAM • BOSTON • HEIDELBERG • LONDON
NEW YORK • OXFORD • PARIS • SAN DIEGO
SAN FRANCISCO • SINGAPORE • SYDNEY • TOKYO
Newnes is an imprint of Elsevier

Newnes is an imprint of Elsevier
30 Corporate Drive, Suite 400, Burlington, MA 01803, USA
Linacre House, Jordan Hill, Oxford OX2 8DP, UK

Copyright © 2006, Elsevier Inc. All rights reserved.

No part of this publication may be reproduced, stored in a retrieval system, or transmitted in any form or by any means, electronic, mechanical, photocopying, recording, or otherwise, without the prior written permission of the publisher.

Permissions may be sought directly from Elsevier's Science & Technology Rights Department in Oxford, UK: phone: (+44) 1865 843830, fax: (+44) 1865 853333, E-mail: permissions@elsevier.com. You may also complete your request on-line via the Elsevier homepage (http://elsevier.com), by selecting "Support & Contact" then "Copyright and Permission" and then "Obtaining Permissions."

Recognizing the importance of preserving what has been written, Elsevier prints its books on acid-free paper whenever possible.

Library of Congress Cataloging-in-Publication Data
Application submitted

British Library Cataloguing-in-Publication Data
A catalogue record for this book is available from the British Library.

ISBN 13: 978-0-7506-7893-3
ISBN: 0-7506-7893-3

For information on all Newnes publications
visit our Web site at www.books.elsevier.com

05 06 07 08 09 10   10 9 8 7 6 5 4 3 2 1

Printed in the United States of America

Working together to grow
libraries in developing countries

www.elsevier.com | www.bookaid.org | www.sabre.org

ELSEVIER   BOOK AID International   Sabre Foundation

*This book is dedicated to my wife, Michela,
and to my children, Lorenzo and Francesco.*
—R. A.

Table of Contents

Preface ix
    by Edmond J. Thomas

Introduction xiii
    by Roberto Aiello

Chapter

1. *History of Ultra Wideband Communication Systems* 1
   by Roberto Aiello

2. *UWB Spectrum and Regulations* 17
   by Robert Sutton

3. *Interference and Coexistence* 53
   by Roberto Aiello

4. *UWB Antennas* 73
   by James S. McLean and Heinrich Foltz

5. *Direct-Sequence UWB* 147
   by Michael McLaughlin

6. *Multiband Approach to UWB* 167
   by Charles Razzell

7. *Spectral Keying^{TM}: A Novel Modulation Scheme for UWB Systems* 191
   by Naiel K. Askar, Susan C. Lin, and David S. Furuno

8. *Multiband OFDM* 211
   by Jaiganesh Balakrishnan and Anuj Batra

9. *MAC Designs for UWB Systems* 249
   by Larry Taylor

10. *Standards for UWB Communications* 285
    by Jason L. Ellis

11. *Commercial Applications* 299
    by Roberto Aiello

About the Contributors 313

Index 317

Foreword
by Edmond J. Thomas

It was Valentine's Day 2002, and I was the chief engineer of the Federal Communications Commission. I sat before an open meeting of the commission and recommended, on behalf of the Office of Engineering and Technology, that it approve a Report & Order authorizing ultra wideband (UWB) for use in the United States. This ended one of the most contentious proceedings in the history of the commission.

The opposition to UWB stemmed from the fact that for the first time in its history, the FCC was considering authorizing a technology that could occupy 7.5 GHz (3.1 to 10.6 GHz) of already licensed spectrum. Although the proposed transmitted power was extremely low (–41 dbm/MHz), the incumbents opposed it with vigor, concerned—needlessly in my view—about the possibility of interference.

Since the proposed spectrum for UWB would cover so much bandwidth if authorized, it also would utilize spectrum already assigned by the National Telecommunications and Information Agency (NTIA) to government users. Some government agencies opposed UWB with the same vigor as the private-sector incumbents. Press articles appeared suggesting that aviation safety, cell phones, GPS, military communications, and satellite communication would be significantly and dangerously impaired by UWB operations. Scientific studies were presented with unrealistic assumptions in support of these kinds of claims. Finally, based on what they were hearing from both government agencies and incumbent private-sector licensees, many members of Congress raised concerns about authorizing UWB.

In spite of all the resistance, on February 14, 2002, the FCC approved the UWB Report & Order, thus giving UWB an opportunity to succeed in the U.S. marketplace. There are too many people to mention both at the FCC and NTIA who deserve credit for supporting and promoting this exciting and innovative technology. However, the following people stand out among the many and deserve mention for their unwavering dedication to get at the truth and do the right thing:

- Michael Powell, former chairman of the FCC
- Kathleen Abernathy, former FCC commissioner
- Kevin Martin, former FCC commissioner, presently FCC chairman

- Michael Copps, FCC commissioner
- Jonathan Adelstien, FCC commissioner
- Bruce Franca, deputy chief of the FCC Office of Engineering and Technology
- Julius Knapp, deputy chief of the FCC Office of Engineering and Technology
- Nancy Victory, former assistant secretary of commerce in charge of NTIA
- Michael Gallagher, assistant secretary of commerce in charge of NTIA

On March 2005, the FCC made another significant decision. It granted a waiver that, in effect, made Multiband Orthogonal Frequency Division Multiplexing (MB-OFDM) UWB technology practical. Today MB-OFDM and direct-sequence impulse are the two primary technologies being commercialized for UWB.

It is not very often that an industry's exact date of birth can be identified. However, UWB's birthday certainly was Valentine's Day 2002. In my view, it was a great Valentine's Day present from the FCC to the nation.

At this writing four years later, the UWB industry is healthy and growing. Today, innovative products are being delivered to U.S. customers, and the technology is beginning to be accepted around the world.

The strength of UWB technology lies in its ability to transfer broadband multimedia content over short distances (100 ft.) efficiently and economically. At the 2006 Consumer Electronics Show in Las Vegas, Nevada, many chip and product manufacturers exhibited and demonstrated products. These companies included Intel, Hewlett-Packard, Motorola, Samsung, Philips, Staccato, Freescale, and Alereon, to name but a few. Demonstrated applications included wireless video streaming, MP3 downloads, multimedia print file transfers, disk backup, and digital camera image transfers.

On the international front, UWB is beginning to be recognized worldwide. On December 8, 2005, ECMA (formerly the European Computer Manufacturer's Association) adopted the MB-OFDM U.S. UWB standard. ECMA is now cooperating with the European Telecommunications Standards Institute (ETSI) to establish a European standard. Also, UWB is on the fast track to becoming an International Standards Organization standard, and the Bluetooth Special Interest Group is moving to adopt UWB as a Bluetooth standard. Finally, MB-OFDM has been accepted as a wireless USB standard. As a result of all this standards activity, it is probable that by the end of 2006, Europe, Korea, and Japan will authorize UWB. China, New Zealand, Australia, Canada, Singapore, and Hong Kong will probably follow shortly there after.

It is estimated that in 2006 over 13 million products will ship containing UWB technology, and if the world market develops as anticipated, over 190 million products will ship in 2009 containing the technology. This corresponds to a growth rate in excess of 140 percent.

In the not too distant future, I believe that UWB chip sets will cost less than $5 (probably around $2). Also, effective operating range will increase to approximately 150 ft. Therefore, in my judgment, UWB will replace today's Bluetooth in applications requiring broadband information transfer and will become the technology of choice for short-range broadband access. It will replace much of today's wiring which connects equipment to computer workstations and laptops, such as printers, monitors, disk drives, scanners, cameras, and the like. It will do the same for components used in entertainment systems, such as tuners, speakers, HDTV monitors, DVD players, and the like. It will also be the technology of choice for personal local area networks. In short, it will revolutionize short-range broadband access.

Dr. Roberto Aiello is uniquely qualified to edit this book and write sections of it. He is a cofounder and CTO at Staccato Communications, and for years he has led the way to making UWB a commercial reality. He was previously founder, president, and CEO at Fantasma Networks, a UWB product company. Prior to his work at Fantasma, in 1996, Dr. Aiello had joined Interval Research, Paul Allen's research laboratory, to work on advanced wireless technologies; there he built the first documented UWB network.

Dr. Aiello is a recognized leader in the UWB community and is actively involved in regulatory and standards-setting committees. He is a founding member of several standards committees, such as the UWB Working Group, the Multiband Coalition, and the Multiband OFDM Alliance, and he now serves as an officer on the WiMedia's board of directors.

Through Dr. Aiello's efforts, in my view, this book will become the definitive UWB reference. In one place it describes the following:

- The history of UWB communication systems
- FCC rules and testing procedures
- Interference considerations
- Antenna designs
- Direct-sequence and MB-OFDM technologies
- Industry standards
- UWB roles and applications

I recommend the book to both engineers and regulators worldwide. In fact, in my judgment it should be required reading for anyone who is interested in an informed and unbiased description of this exciting new technology. In short, it establishes the reality and destroys the myths surrounding UWB. I congratulate Dr. Aiello and his cocontributors for a job well done.

Edmond J. Thomas
Former Chief Engineer,
Federal Communications Commission
Technical Policy Advisor and Partner
Harris Wiltshire Grannis
ethomas@harriswiltshire.com

Introduction
by Roberto Aiello

I discovered Ultra Wideband (UWB) in 1996 when I joined Interval Research, a research and technology incubator company funded by Paul Allen, to lead their wireless activities. At the time, we were exploring the opportunity to enable connectivity of personal and wearable devices, such as video cameras, stereo headsets, head-mounted displays, personal storage, and so forth.

We needed a high-performance, short-range, low-power wireless network. We studied the options of using spectrum in the 2.4 GHz and 900 MHz bands, infrared, or inductive coupling. Each has its advantages, but none of them could solve our problem. Then, I read an article about impulse radios published by Time Domain, and I decided to explore that opportunity.

As I was introduced to the world of UWB, I had the privilege to meet and collaborate with many UWB pioneers whose contributions made UWB a reality today: Bob Fleming and Cherie Kushner at Aetherwire, Bob Fontana at Multispectral Solutions, Larry Fullerton and Paul Withington at Time Domain, Martin Rofheart and John McCorkle at XtremeSpectrum, Bob Scholtz at University of Southern California. All of these individuals were driven by different interests: to work on new engineering development, to research novel communications areas, or to create a new business. Their passion and dedication have greatly contributed to the current success of UWB.

Even though UWB research dates far back to the beginning of radios, it was reborn as a result of the Federal Communications Commission (FCC) Report and Order in 2002, when spectrum was made available for unlicensed use, and companies recognized a business opportunity. This resulted from the effort of a lot of people who actively promoted UWB at the FCC and in the government.

One of the biggest contributions, in my opinion, was Jim Lovette's. With his endless energy and dedication, he worked extremely hard and efficiently to explain what UWB was about to both engineers and policy makers. Jim was a self-defined "unlicensed spectrum advocate"; he was the originator of the Apple petition that led to the unlicensed national information infrastructure (UNII) band spectrum allocation in the 5 GHz band and unlicensed personal communication services (PCS) in the 1.9 GHz band. We worked together for 18 months to lobby the FCC, National

Telecommunications and Information Agency (NTIA), Congress and Senate to allocate UWB spectrum in the United States. He and I had several heated discussions during that time, and I obviously had different opinions in many situations, but in hindsight, I must admit that he was right in most cases! He passed away on June 29, 2002.

The FCC was a great supporter of UWB from the beginning, and the tireless work of Julie Knapp, Dale Hatfield, and Ed Thomas greatly contributed to the final UWB spectrum allocation.

After 2002, the scenario changed, and a push began to capitalize on the UWB spectrum available and to develop commercial opportunities. Today, UWB is one of the hottest wireless technologies in the industry.

The purpose of this book is to introduce the reader to high-performance UWB communication systems and to UWB's commercial applications. I did not include other applications, such as low-bit-rate communications, radar, and radio frequency (RF) tags, among other others. Nor did I include a chapter on impulse radios, mainly because, even though it was one of the main methods developed for UWB, it has not found a place in high-performance communication applications. Furthermore, quite a few publications and books have been written on the topic.

The book is logically divided into three parts: introduction to UWB (Chapters 1–4), different UWB techniques (Chapters 5–8), and communication applications (Chapters 9–11).

The main theme of the book is that UWB is not a specific technology, modulation, or multiple access technique; nor is it a specific application. Rather, it is defined as available spectrum that needs to be used according to specific rules. This spectrum is characterized by transmitters that emit very low average power and, as such, generate a very low level of interference with other systems sharing the same spectrum. This enables the unique property of UWB, which is spectrum open to many services that do not cause harmful interference with each other. This results in a new and more efficient use of the spectrum.

Several technologies have been developed to use the UWB spectrum so far. The more successful commercially are based on well-known communication techniques, such as direct sequence spread spectrum or orthogonal frequency division multiplexing (OFDM). They also differentiate each other by using a single band or multiple bands in the same system.

The main applications developed to date, allowing the best use of the UWB spectrum, focus on two extreme cases: (1) very high bit rate, low power, and short range, and (2) very low power, low bit rate. Useful features of such systems are also location and positioning. These applications have emerged because their requirements are a good fit for UWB and because they are not currently served by other wireless systems. As mentioned above, this book focuses on high-bit-rate systems.

Industry standards have been developed to support commercial applications and to facilitate interoperability among multiple vendors. Such standards are also described in the book.

The authors are all "A" players in the space and have been key contributors to the development of modern UWB and to the technologies that have moved UWB from the laboratory to commercial products.

This book is organized as follows:

- *Chapter 1: History of Ultra Wideband Communication Systems.* UWB was originally developed as impulse radio, using its characteristics for low probability of intercept and low probability of detection for military applications. This chapter describes it origins and the motivations that led to the development of UWB systems. It summarizes the history of UWB, focusing on the more recent events that led to the pursuit of commercial applications.

- *Chapter 2: UWB Spectrum and Regulations.* UWB is defined as available spectrum. As of today, it has been regulated in the United States, while the rest of the world is still working toward a global spectrum allocation. UWB rules are different from any other existing spectrum regulations, and UWB transmitters need to meet very stringent requirements to be allowed to operate. This chapter describes the FCC rules and the measurement procedure required to test a UWB transmitter for compliance.

- *Chapter 3: Interference and Coexistence.* UWB systems need to coexist with other systems sharing the same frequency spectrum. They also use a very large bandwidth to compensate for the very low allowed transmit power. This chapter describes some of the issues related to interference and coexistence.

- *Chapter 4: UWB Antennas.* The subject of antennas has great importance for UWB technology: a system that employs very large bandwidth presents challenges to the antenna designer. This chapter describes some of these challenges and some of the antennas that are adequate for commercial applications, including design details and an explanation of their differences from conventional narrow-channel antennas.

- *Chapter 5: Direct Sequence UWB.* This chapter describes a direct sequence UWB system (the most popular version of the original impulse radios for high-performance systems), its characteristics, and its advantages.

- *Chapter 6: Multiband Approach to UWB.* A multiband system is based on the principle of transmitting different symbols in different frequency bands in a periodic sequence. This chapter describes the general principles that led to the development of multiband techniques, their advantages, and their characteristics.

- *Chapter 7: Spectral Keying: A Novel Modulation Scheme for UWB Systems.* This chapter describes a modulation scheme based on multibands, which uses the relationship between bands to encode information.

- *Chapter 8: Multiband OFDM.* This chapter describes a system based on multiband and OFDM that has become attractive in terms of cost-performance ratio for high-bit-rate, short-range systems.

- *Chapter 9: MAC Designs for UWB Systems.* UWB systems are characterized by unique constraints, different from other wireless communication systems. This chapter describes MAC architecture optimized for UWB systems.
- *Chapter 10: Standards for UWB Communications.* Industry standards are necessary for the successful development of wireless systems. This chapter describes the status of the ongoing standard effort related to UWB systems.
- *Chapter 11: Commercial Applications.* The success of wireless products depends on several factors: technical characteristics, target applications, market timing, and economics. This chapter describes how these various aspects are relevant to UWB products.

This book is intended for students, engineers, and marketing and business people, because it covers a broad set of fundamental issues related to UWB.

It has been a pleasure to work on this book with so many friend and colleagues. Many individuals have helped with support, advice, discussion, comments, and constructive criticism: Mark Bowles, Billy Brackenridge, Jeff Foerster, Anita Giani, Harry Helms, Kursat Kimyacioglu, Dave Leeper, Janine Love, Sid Shetty, Jarvis Tou, and Stephen Wood. I am most in debt to the individuals who encouraged me to continue working on UWB for so many years: most shared my drive to create something new (a technology or business), and some were motivated by the prospect of financial return (Venture Capitalists that invested in my companies).

But most importantly, I couldn't have been successful in pursuing my interest in technology and in fulfilling my entrepreneurial drive without the support of my wife of sixteen years, Michela, who has shown an amazing amount of patience and continuous encouragement over the years.

1

History of Ultra Wideband Communication Systems

by Roberto Aiello

From its humble beginnings more than 45 years ago, ultra wideband (UWB) technology has traveled an interesting road from the lab, to the military, back to the lab, and finally into commercial prototyping and implementation. Known throughout the years by a number of different names, fundamentally, UWB offers a different mechanism for wirelessly transporting voice, video, and data. And, due to its broad bandwidth, it enables both high-data-rate personal-area network (PAN) wireless connectivity and longer-range, low-data-rate applications.

The term *ultra wideband* was coined in the late 1980s, apparently by the U.S. Department of Defense [1], and the actual technology behind UWB has been known by many other names throughout its history, including baseband communication, carrier free communication, impulse radio, large relative bandwidth communication, nonsinusoidal communication, orthogonal functions, sequency theory, time domain, video-pulse transmission, and Walsh waves communication [2].

More recently, with the UWB spectrum allocation in the United States, a better definition for UWB is "available spectrum," independent of the type of technique used to transmit the signal. In this case, it is imperative that transmissions over the UWB spectrum be able to coexist with other services without causing harmful interference.

The credit for developing UWB technology belongs to many innovative thinkers and scientists over the last 50 years. Interest in the technology has been steady, with more than 200 technical papers published in journals between 1960 and 1999 on the topic and more than 100 U.S. patents issued on UWB or UWB-related technology [3]. It should be noted that, given the lack of spectrum regulations, the early work was performed without specific commercial applications in mind, which would eventually become necessary for successful product deployment.

After decades of development, UWB is now most compelling in that its standards are in place and the timing is finally right for this enabling technology to achieve mass adoption. It is particularly interesting that early work on pulse-based transmission led to the UWB spectrum allocation. In the United States, UWB spectrum allocation allowed development of commercial products that led to non-pulse-based transmission techniques.

1.1 Understanding UWB

It is generally agreed that the origins of UWB trace back to some seminal development work in the late 1950s and early 1960s, spearheaded by Henning Harmuth and Gerald Ross. However, as early as 1942, Louis de Rosa filed for a patent that took a different approach from the "traditional" sinusoidal radio waves made ubiquitous by the work of Guglielmo Marconi. He was granted U.S. Patent No. 2671896 for a random impulse system in 1952 "to provide a method and means to generate a series of pulses of random occurrence which, when used for modulation of a carrier for transmission purposes, greatly minimizes the possibility of enemy jamming" [4]. In 1945, Conrad H. Hoeppner filed for another UWB-related patent (which was granted in 1961) for a pulse communication system that reduces interference and jamming [4].

One of the early and continuing motivations to work with a very large bandwidth is that it provides significant advantages to bit rate and power consumption.

In terms of low power, UWB maximum output power currently allowed by Federal Communications Commission (FCC) regulations is 0.0001 mW/MHz. This compares favorably with the 100 mW maximum output power of the IEEE 802.11b specification for wireless local area networks (WLANs) or even the 40 mW output power of the IEEE 802.11a specification. It should be noted that the UWB power is given in terms of power spectral density, not total power. The maximum total power of a UWB system occupying a 1 GHz bandwidth, for example, would be 0.1W, still much lower than other comparable radios. If the system is designed correctly, this can lead to a UWB implementation with low power consumption and low interference to other systems sharing the same frequency of operation.

In terms of bit rate, UWB reaps the benefits of a broad spectrum. According to the following formula [5], where C = capacity, W = bandwidth, and S/N = signal-to-noise ratio,

$$C = W \cdot \log_2\left(1 + \frac{S}{N}\right) \qquad (1.1)$$

Obviously, it is easier to increase the bit rate (capacity) by increasing the bandwidth instead of the power, given the linear-versus-logarithmic relationship. As an example, assuming S/N = 12 dB, it would be necessary to increase either the bandwidth by a factor of two or the power by a factor of four in order to double the capacity. For this reason communication engineers prefer to increase the system bandwidth instead of the power to achieve higher bit rates. This means that systems that use the UWB spectrum can be designed to achieve high bit rate better than narrow bandwidth systems.

To double the range instead, according to the following Friis formula, where d = distance, P_t = transmit power, and P_r = receive power,

$$d \propto \sqrt{\frac{P_t}{P_r}} \qquad (1.2)$$

it would be necessary to increase either the bandwidth or power ratio by a factor of four.

So, in summary, it is more efficient to achieve higher capacity by increasing bandwidth instead of power, while it is equally difficult to achieve a longer range. Thus, UWB designers focused on higher-bit-rate, short-range systems initially.

The successful development of UWB was primarily dependent on advancements in three key areas: theory, tools, and practice.

1.2 Theory

In the late 1950s, work was being performed at Lincoln Laboratory in Lexington, Massachusetts, and Sperry Research Center (SRC) in Sudbury, Massachusetts, to develop phased-array radar systems. Specifically, engineers at Sperry developed an electronic scanning radar (ESR) that connected 3 dB branch line couplers in order to form a two N-port network. Each input port corresponded to a particular phase taper across the output N-ports which, when connected to antenna elements, corresponded to a particular direction in space. This ESR device was known as a Butler Hybrid Phasing Matrix [3].

While characterizing the wideband properties of this network, the team worked to reference the properties of the four-port interconnection of quarter-wave transverse electromagnetic wave (TEM) mode lines, which formed the branch line coupler. The analysis began by studying the impulse response of the networks [3].

In addition to the work of Ross and Kenneth W. Robbins at what became Sperry Rand Corporation and later Sperry Corp, much of the early work behind UWB theory was performed under the guidance of Harmuth at Catholic University of America and Paul Van Etten at the U.S. Air Force's Rome Air Development Center [6]. From 1960 on, papers began appearing in scientific journals exploring the idea of electromagnetic radio signals without a sinusoidal time variation.

Beginning in 1969, with *Transmission of Information by Orthogonal Functions*, until 1984, Harmuth published a series of books and papers that outlined the basic design for UWB transmitters and receivers, and he was awarded numerous patents. In the 1970s, Van Etten performed extensive empirical testing of UWB radar systems, which led to his development of system-design and antenna concepts.

Though not first in line to develop UWB theory, the U.S. government was reasonably quick to recognize its potential. In the 1960s, both Lawrence Livermore National Laboratory (LLNL) and Los Alamos National Laboratory researched pulse transmitters, receivers, and antennas. In the 1970s, LLNL expanded its laser-based diagnostics research into pulse diagnostics. LLNL was granted numerous patents in the 1990s for its work in UWB.

During this period, much of the fundamental theory of what became UWB was worked out, as engineers and scientists sought to describe and characterize the behavior of microwave networks through the characteristic impulse of their radiating elements in the time domain. This became known as time-domain electromagnetics.

Traditionally, a linear, time-varying system was characterized by taking amplitude and phase measurements versus frequency, also known as a swept frequency response. Now the system could be analyzed through its response to an impulse excitation, also known as an impulse response $h(t)$. Specifically, the output $y(t)$ of a UWB system to any arbitrary input $x(t)$ could be determined using the convolution integral [7]:

$$y(t) = \int_{-\infty}^{\infty} h(u) x(t-u) du$$

By 1965, interest in time-domain electromagnetics was growing. Some key areas of research included

- J. Lamar Allen: analysis in linear and nonreciprocal microwave networks and antennas to ferrite devices
- Harry Cronson: time-domain metrology, studying the frequency-domain properties of passive microwave networks through impulse response and Fourier transforms
- David Lamensdorf and Leon Susman: analysis of antennas using time-domain techniques
- C. Leonard Bennett and Joseph D. DeLorenzo: studying the impulse response of targets directly in the time domain
- Joseph D. DeLorenzo: invention of the time-domain scattering range for scattering analyses of targets and antennas
- Ohio State: continued study of scattering, using individual frequencies and combining the amplitude and phase data by computer using DeLorenzo's scattering-range concepts [3]

By 1977, much of the theory regarding nonsinusoidal waves was coming together, and Harmuth published the work *Sequency Theory*, which included a chapter entitled "Electromagnetic Waves with General Time Variation." This chapter covered practical radiators, receivers, and applications for radar, including, for example, a correlation receiver for selective reception of nonsinusoidal waves.

Fourteen years later, Harmuth published *Nonsinusoidal Waves for Radar and Radio Communication*. By this time, using a pulse compressor for nonsinusoidal waves was generally recognized as analogous to the tuned resonant circuit for the selective reception of sinusoidal waves.

During the late 1970s and 1980s, there was also a number active academic research programs in UWB, including the University of Michigan, the University of Rochester, and Brooklyn University [6].

1.3 Tools and Techniques

While the work in UWB theory was demonstrating that studying responses in the time domain was the right approach, measuring them was an entirely different matter. Barney Oliver at Hewlett-Packard broke the first logjam in 1962 with the development of the sampling oscilloscope. When combined with the technique of using avalanche transistors and tunnel diodes to generate very short pulses, the oscilloscope made it possible to observe and measure the impulse response of microwave networks directly.

In the late 1960s, Tektronix developed commercial sample-and-hold receivers. Although not designed for UWB, these receivers used a technique that could be used to enable UWB signal averaging. (The sampling circuit is a transmission gate followed by a short-term integrator [6].)

Once these impulse-measurement techniques were applied to the design of wideband radiating antenna elements, the logical next step was to begin developing short-pulse radar and communications systems with the same set of tools. As a further advancement, Robbins invented a sensitive, short-pulse receiver (U.S. Patent No. 3,662,316; 1972) that could be used in place of the time-domain sampling oscilloscope [7].

The final enabling item for UWB development was the threshold receiver. Soon after the advent of the oscilloscope, avalanche transistors and tunnel diode detectors were used to detect short duration signals (in the neighborhood of 100 ps). Eventually, A. Murray Nicolson developed the tunnel diode constant false alarm rate (CFAR) receiver. A type of CFAR detector is still in use [3].

Also during the 1970s, Ross and Nicolson developed a narrow baseband pulse fixture combined with fast Fourier transform (FFT) techniques to measure the stealthy properties of microwave absorbing materials from a single pulse measurement. Later, Hewlett-Packard developed the network analyzer to provide this same functionality.

By the end of the 1970s, then, the basic tools and techniques were in place to begin designing UWB signal systems, and the fundamental system theory was established. Key major components included pulse train generators, pulse train modulators, switching pulse train generators, detection receivers, and wideband antennas. Significant subcomponents and techniques included avalanche transistor switches, light responsive switches, the use of "subcarriers" in coding pulse trains, leading edge detectors, ring demodulators, monostable multivibrator detectors, integration and averaging matched filters, template signal match detectors, correlation detectors, signal integrators, synchronous detectors, and antennas driven by stepped amplitude input [6].

1.4 Practice

In a paper for the Institute of Electrical and Electronics Engineers (IEEE) published in 1978, Bennett and Ross presented examples of the use of UWB technology in radar and communications applications, including collision avoidance systems, spacecraft docking, airport surface-traffic control, auto braking, ship docking, liquid-level sensing, and wireless communications [8].

During the 1970s and 1980s, practical development work continued, and many patents were granted in the field of UWB. Patents, rather than articles and publications, are the best way to track progress during this period of time. For instance, Ross was granted a patent (U.S. Patent No. 3,728,632; 1973) titled "Transmission and Reception System for Generating and Receiving Base-Band Pulse Duration Pulse Signals without Distortion for Short Base-Band Communication System." The same year, Robbins was granted a patent for a "base-band transmitter and receiver antenna system for operation in sub nanosecond pulse radio systems" (U.S. Patent No. 3,739,392; 1973).

With an early commercial success, Rexford Morey, working at the Geophysical Survey Systems (North Billerica, Massachusetts), was granted a patent for a geophysical survey system that determines the character of the subterrain by analyzing the reflections from 1 ns electromagnetic pulses radiated into the ground (U.S. Patent No. 3,806,795; 1974). This was followed by a patent for Jamie C. Chapman filed in 1974 for "signal processing techniques and apparatus for use in short-pulse geophysical radar systems to improve signal-to-noise ratio, reduce radio frequency (RF) interference, improve resolution, and reduce ambiguities" (U.S. Patent No. 4,008,469; 1977).

In 1978, Harry Cronson and his colleagues from Sperry filed for a patent for a "collision avoidance system using short-pulse signal reflectometry" (U.S. Patent No. 4,254,418; 1981). In 1983, Harmuth began the patent process for a "frequency independent shielded loop antenna" designed for radiating and receiving nonsinusoidal electromagnetic waves (U.S. Patent No. 4,506,267; 1985).

In all, Sperry was awarded more than 50 patents for UWB pulse-generation and -reception methods by 1989, covering applications such as communications, radar, automobile collision avoidance, positioning systems, liquid-level sensing, and altimetry [7]. By the late 1970s, interest began growing in using UWB for voice and data communications systems. Ross reports some early work in this area. Robert Fontana, who became president of Multispectral Solutions (Germantown, Maryland), performed much of the early work in this area as well and continues to be a leader in the development of advanced UWB systems for communications, radar, and geopositioning.

The U.S. government and other national governments were also active in UWB system development during this period. Work in the commercial and military sectors came together, for example, when Ross and Fontana developed a low probability of intercept and detection (LPI/D) communications system for the U.S. government in 1986. The two continued their collaboration efforts for UWB radar and communications systems until 1997 [7].

According to many, much of the early work in the UWB field (prior to 1994), particularly in the area of impulse communications, was performed under classified U.S. government programs. More recently, however, much of the work has been carried out without classification restrictions; as a result, the development of UWB technology has greatly accelerated.

In 1992, the U.S. government was awarded a patent for work done by Robert E. Jehle and David F. Hudson for an "impulse transmitter and quantum detection radar system." In this system, "pulses of short duration are radiated by an impulse radar transmitter with time-domain intervals under control of a clock to monitor a moving target from which echo pulses are reflected. The echo pulses received are measured by a quantum detector in terms of photon energy levels to supply signal data processed during periods between said time-domain intervals" (U.S. Patent No. 5,095,312; 1992).

Work in radar applications for UWB continued throughout the 1990s. As an employee at Lawrence Livermore National Laboratory, T. E. McEwan invented micropower impulse radar (MIR) in 1994, which featured very low power and was extremely compact and relatively inexpensive. Work also continued in the commercial sector for applications like security systems.

Larry Fullerton at Time Domain (Huntsville, Alabama) developed a communication system based on pulse-position modulation [9]. This system, although expensive to build for commercial applications, initially generated a lot of interest within the industry. The promotional work of Paul Withington and Ralph Petroff at Time Domain had also caused the general public to identify pulse-position modulation as the main type of UWB and generated interest among the public and within academia and the government.

The first development that targeted consumer products was started by the author in 1996 at Interval Research, a research company based in Palo Alto, California, funded by Paul Allen, one of the cofounders of Microsoft [10]. The goal was to develop low-power, high-performance wireless networks for Personal Area Networks (PANs). Product concepts considered included belt-to-ear stereo links, video links for personal media players, video links from cameras to recording units for wearable cameras. The team at Interval Research studied the feasibility of various transmission techniques in the Industrial Scientific and Medical (ISM) bands (2.4 GHz) and the Unlicensed National Information Infrastructure (U-NII) bands (5 GHz), and it also studied inductive coupling. The research revealed that none of these methods would meet the application requirements, so the team decided to develop a system based on UWB.

The result of this work was a commercial network architecture, including UWB transmitters (U.S. Patent No. 6,658,053), receivers (U.S. Patent No. 6,275,544), and low-cost antennas (U.S. Patent No. 6,292,153). The team at Interval Research also actively lobbied the FCC and the U.S. government to allocate UWB spectrum in the United States. The technology was eventually used to form Fantasma Networks, and several key members of this original team later joined Staccato Communications.

XtremeSpectrum was the other company that began pursuing commercial opportunities in 1999. The two founders, Martin Rofheart and John McCorkle, developed a high-pulse-repetition frequency technology different from any other system at the time. The company eventually was acquired by Freescale in 2004.

1.5 UWB in the Twenty-first Century

Even though the 1980s and 1990s were marked by UWB companies' receiving generous funding that resulted in the creation of successful applications and persistent governmental lobbying, neither ever led to commercial acceptance.

Most notably, UWB lacked three crucial boosters:

- Spectrum allocation by the FCC
- An industry standard
- Simple, easy implementation

1.5.1 Spectrum

The FCC made UWB a reality with its Report & Order of February 14, 2002 [11]. Until then, UWB was thought of as technology that operates by transmitting very short pulses in rapid succession, as described above. The FCC's definition of the criteria for devices operating in the UWB spectrum purposely did not specify the techniques related to the generation and detection of RF energy; rather, it mandated compliance with emission limits that would enable coexistence and minimize the threat of harmful interference with legacy systems, thus protecting the Global Positioning System (GPS), satellite receivers, cellular systems, and others.

Today, UWB, according to the FCC, is defined as any radio transmitter with a spectrum that occupies more than 20 percent of the center frequency or a minimum of 500 MHz and that meets the power limits assigned by the FCC. Recognizing the advantages of new products that incorporated this technology for public-safety, enterprise, and consumer applications, the FCC's Report & Order allocated unlicensed radio spectrum from 3.1 GHz to 10.6 GHz expressly for these purposes. The move, which legitimized the UWB market, also fueled the urgency behind creating and accepting an industry standard.

Many companies and organizations have contributed to the effort to obtain UWB spectrum. Special credit goes to Aetherwire and Location, Fantasma Networks, Interval Research, Multispectral Solution, Time Domain, and XtremeSpectrum. These companies have led the lobbying effort and contributed technically to help the FCC identify the best solution to allow UWB applications and protect incumbent services.

1.5.2 Developing an Industry Standard

While the UWB spectrum could be used by impulse radios built during previous generations, a movement soon began to create an updated approach that would best maximize the efficiency and technical prowess needed for current and next generation high-performance commercial applications.

Industry leaders recognized that the adoption of a standard was key to the successful widespread implementation of UWB. Standards organizations, such as the the Institute of Electrical and Electronics Engineers (IEEE) and Ecma International, and industry groups, including the USB Implementers Forum and the WiMedia Alliance, were dedicated to ensuring that UWB met FCC guidelines and delivered performance far beyond that achieved by other wireless technologies, such as 802.11 and Bluetooth, to meet new application requirements.

1.5.2.1 Multiband

To further fuel industry momentum, in October 2002 four companies, General Atomics, Intel, Staccato Communications (then called Discrete Time Communications), and Time Domain (replaced by Alereon), created the Multiband Coalition. Philips and Wisair joined in January 2003. Banded together by a common belief that the UWB standard should be based on "multiband" technology, the coalition worked to refine its united vision, create a common draft specification, and repeatedly test its performance. The concept of multiband is to break the available spectrum into subbands (each at least 500 MHz wide because of the FCC ruling) and to communicate in those independently. Many individuals from different companies came to the conclusion that this would be a superior technique for commercial applications, but the first to develop the concept was probably Gerald Rogerson at General Atomics.

In January 2003, before a group of major consumer electronics makers, including Eastman Kodak, Panasonic, Samsung Electronics, and Sony, along with leading integrated circuit designers, including Infineon Technologies AG and STMicroelectronics NV, the Multiband Coalition put on a demonstration that resulted in additional backers and generated serious interest.

Moving forward, the coalition members recognized that their audience shared a common goal: securing an industry standard that would help produce the best possible physical layer (PHY) specification. Out of that gathering came further cooperation, performance testing, and development efforts from industry participants, with additional members joining the coalition.

In the process, the original vision took an unexpected turn. At the IEEE meeting in March 2003, Texas Instruments presented a proposal based on Orthogonal Frequency Division Multiplexing (OFDM) [12] that showed available range performance superior to all other proposals at the time.

Some members of the coalition collaborated with Texas Instruments to validate their results. As part of that effort, Staccato duplicated Texas Instruments' system simulation and independently analyzed the system's complexity, presenting the results at the IEEE meeting in May 2003 [13]. Further discussions followed, involving all coalition members and, finally, at a meeting in Denver, Colorado, and a follow-up in San Diego, California, hosted by General Atomics, in June 2003, the decision was made to adopt OFDM as the preferred modulation scheme for multiband.

In order to better reflect the evolution of the multiband movement, the Multiband Coalition was reborn as the Multiband OFDM Alliance (MBOA) [14] in June 2003. At that time, Hewlett-Packard, Microsoft, NEC Electronics, Panasonic, Samsung, Sony, STMicroelectronics, SVC Wireless, TDK, Texas Instruments, and Wisme joined existing coalition members Femto Devices, Focus Enhancements, Fujitsu, General Atomics, Infineon, Institute for Infocomm Research, Intel, Mitsubishi Electric, Royal Philips Electronics NV, Staccato Communications, Taiyo Yuden, Time Domain, and Wisair.

1.5.2.2 Multiband OFDM Technology

The basic concept behind multiband OFDM divides spectrum into several 528 MHz bands (with each occupying more than 500 MHz at all times in order to comply with FCC regulations).

Information is transmitted via 128 subcarriers in each band using quadrature phase-shift keying (QPSK) modulation. With a contiguous set of orthogonal carriers, the transmit spectrum always occupies a bandwidth greater than 500 MHz.

This method takes advantage of frequency diversity while delivering necessary robustness against interference and multipath, even in the most challenging channel environments. The signal is sequenced across three or seven bands, with 9.5 ns switching time.

A multiband OFDM system's radio architecture is similar to other conventional OFDM systems. This is an advantage because OFDM has been widely adopted by other standards organizations for applications that require high performance communication systems, including ADSL, 802.11a/g, 802.16a, digital audio broadcast, and digital terrestrial television broadcast.

In a simple configuration, necessary data rates (110 Mbps and 200 Mbps) need a single digital analog converter (DAC) and mixer for the transmit chain for reduced complexity. The digital section can scale with future complementary metal oxide semiconductor (CMOS) process improvements as implementers trade simplicity for performance. Power scaling requires a half-rate pulse-repetition-frequency (PRF) approach, which can increase off time to enable power conservation. Implementers can also trade power consumption for range and information data rate. (For more information on multiband OFDM functionality, see Chapter 8.)

1.5.2.3 Settling on a Standard

The IEEE's 802.15.3a Task Group was established in January 2003 to define specifications for high-speed UWB, and it was terminated in January 2006 because it couldn't converge to a specification. While the multiband OFDM proposal earned majority votes, it did not garner the three-quarters approval necessary. An alternative version of UWB based on impulse radio and called direct-sequence UWB (DS-UWB), offered by Freescale Semiconductor, was proposed.

In light of this inability to decide, the industry moved ahead. The MBOA grew to more than 170 members, which included many of the most influential players in consumer electronics, PCs, home entertainment, mobile phones, semiconductors, and digital imaging.

On November 10, 2004, the Multiband OFDM Alliance Special Interest Group (MBOA-SIG) announced that it had completed its PHY 1.0 specifications and made them available to MBOA-SIG Promoter, Contributor, and Adopter members. This move enabled MBOA members to finalize their UWB chip and board-level designs.

In March 2005, WiMedia and the MBOA-SIG, the two leading organizations creating UWB industry specifications and certification programs for consumer electronics, mobile, and PC applications, merged into the WiMedia Alliance. This nonprofit industry association is committed to developing and administering specifications collaboratively from the physical layer up, enabling connectivity and interoperability for multiple industry-based protocols sharing the MBOA-UWB spectrum. To date, the WiMedia Alliance has built a complete ecosystem of silicon, software, hardware, and test equipment to allow the industry to develop solutions.

In February 2005, the Wireless USB Forum announced its support to the WiMedia-MBOA solution, and in June, the Wireless USB Promoter's Group announced the completion of the Certified Wireless USB (WUSB) specification. In April, the 1394 Trade Association followed suit. Since then, the Bluetooth Special Interest Group (BSIG) announced its intent to work with the developers of UWB to combine the strengths of both technologies.

1.6 Simple, Cost-effective Implementation

Some of the first UWB radios have been specifically designed to enable short design cycles when adding UWB capability to devices such as cellular phones, PDAs, and other portable devices.

Quite a few companies have demonstrated prototypes to date, and shipments in consumer products are expected to begin in 2006.

1.6.1 Alereon

Alereon is a fabless semiconductor company providing complete solutions for WUSB/WiMedia UWB applications. Alereon's first chipset (shown in Figure 1.1),

Figure 1.1 Alereon AL4000 family of products.

the AL4000 WiMedia PHY, combines the AL4100 Analog Front End (AFE) and the AL4200 BaseBand Processor (BBP) for a complete WiMedia PHY solution. The AL4100 AFE, in combination with the AL4300 single-chip WUSB media access controller (MAC) and BBP, combines for a complete WUSB device solution.

1.6.2 Staccato Communications

Staccato Communications, based in San Diego, is a UWB radio technology pioneer with application expertise in WUSB, Bluetooth, and Internet Protocol (IP) connectivity. The fabless semiconductor company serves the mobile phone, personal computing, and consumer electronics industries with small form factor, single-chip, all-CMOS, wireless systems-in-packages (SiPs). Staccato's offerings include PHY, MAC, drivers, application software, development kits, reference designs, and support services. Founded by wireless industry veterans who have delivered commercially successful solutions for mobile phones, Bluetooth, and WiFi, Staccato is now leveraging that expertise for UWB-based applications.

Staccato's flagship product, the Ripcord W-USB Series (shown in Figure 1.2), combines a WiMedia-compliant PHY, MAC, 32 bit RISC CPU, buffer memory, host interfaces (USB and SDIO), transmit/receive switch, high-accuracy crystal, UWB filter, RF matching network, and all necessary passive components in a single-chip SiP. Only an antenna and power are required for a complete WUSB solution.

Figure 1.2 Staccato's Ripcord W-USB product.

Figure 1.3 Wisair DV9100 module.

1.6.3 Wisair

Founded in 2001 and located in Tel Aviv, Israel, Wisair has developed a UWB chipset that enables low-cost, low-power, and high-bit-rate communications modules and system solutions for the fast emerging market for home/office connectivity for video/audio and data applications. The Wisair DV9110 module (see Figure 1.3) is based on a two-chip solution, using a SiGe RF transceiver and a CMOS baseband chip. It offers multiple data interfaces and protocols, including Ethernet 100 Mbps, USB 2.0, RMII, MII, Direct CPU interface, and MPEG transport stream SPI.

These products are targeted to consumer applications and are high performing and inexpensive. We can expect even higher levels of integration in the near future, following the same type of evolution the Bluetooth and WiFi markets have followed. This means more that products with more capabilities and lower form factors will be developed.

1.7 The UWB Market

UWB technology will appeal to consumers by enabling an evolution to high-performance WUSB, Wireless 1394, and IP over UWB connectivity solutions. The technology is being designed into computers and peripherals, as well as home entertainment equipment, mobile phones, and other consumer electronics devices. The wireless extensions to the already popular wired protocols are now creating new usage models and adding a convenience that consumers have been demanding—the removal of the unsightly "rat's nest" of cables—while providing ease of installation and true mobile connectivity.

With an industry standard in place, UWB component makers have begun the process of producing large volumes of chips for consumer electronics applications. According to market-research firm Allied Business Intelligence, UWB electronics and chips will reach 45.1 million units by 2007 with anticipated industry revenues of $1.39 billion.

Initial applications are set to include desktop and notebook personal computers, PDAs and handheld computers, mobile phones, printers, scanners, portable audio, multimedia players, external hard drives, digital cameras, digital camcorders, digital TVs, advanced set-top boxes, personal video recorders, and DVD players.

1.8 High-Volume Commercial Applications

The first major application of UWB will be Certified WUSB based on WiMedia UWB radios designed specifically to provide high-speed, short-range, low-power connectivity. Certified WUSB is the wireless evolution of its wired cousin, USB 2.0, and just like wired USB, it is optimized for connecting devices to PCs and other mobile devices. There are nearly 2.5 billion USB connections already in existence worldwide, and more than 600 million USB-enabled PCs shipped in 2005. In short, USB is the most successful connection interface ever, and it is now expanding its appeal beyond the PC to the consumer electronics and mobile phone markets, which represent the majority of growth in USB today. Certified WUSB maintains all of the features that originally made USB successful: ease of use, blazing speed (480 Mbps), and low cost. Far from being competitors, WUSB and WiFi, just like Ethernet and USB in the wired world, complement one another.

The first Certified WUSB products will be dongles and add-in cards to retrofit existing PCs and peripherals. The next step in the evolution will include devices with Certified WUSB embedded directly into them, such as printers. External storage devices are becoming popular in the home because they are a great place to store movies and TV shows captured by PCs with TV tuners. These storage devices can take advantage of the high-bandwidth connection provided by UWB.

Digital still cameras are also a natural fit for WUSB. To date, cameras have just begun featuring digital radios. While it has low power consumption, Bluetooth is too slow to transfer digital photographs, let alone digital video. WUSB, on the other hand, makes it very easy for both digital still cameras and digital video cameras to download or upload pictures and video to PCs. But that is just the beginning. Digital radios should make their way into all consumer electronics products in the near future. It is important that these radios are able to talk to PCs, and WUSB is a great way to ensure that.

It will soon be possible to pick up a UWB-enabled digital video camera, take some videos, and then walk up to any UWB-equipped TV set and play the movie—with no wires.

With yearly shipments of Bluetooth and WiFi devices topping hundreds of millions, UWB backers are eager to greet the mass market for communications and computer systems with a portfolio of products offering the kind of high-bandwidth and low-power capabilities that consumers have yet to appreciate—but certainly will.

References

1. Barrett, T. W. "History of Ultra Wideband (UWB) Radar and Communications: Pioneers and Innovators." *Proceedings of Progress in Electromagnetics Symposium 2000* (PIERS2000), Cambridge, MA, July 2000.
2. Harmuth, Henning F. "Application of Walsh Functions in Communications," *IEEE Spectrum* (November 1969): 82–91.
3. Ross, Gerald F. "Early Motivations and History of Ultra Wideband Technology." White paper, Anro Engineering.
4. See www.aetherwire.com/CDROM/General/patabsmainframe.html (UWB patent listing). Accessed April 2006.
5. Shannon, C. E. "A Mathematical Theory of Communication." *The Bell System Technical Journal* 27 (July and October 1948): 379–423 and 623–656.
6. Barrett, T. W. "History of Ultra Wideband (UWB) Radar and Communications: Pioneers and Innovators." *Proceedings of Progress in Electromagnetics Symposium 2000* (PIERS2000), Cambridge, MA, July 2000.
7. Fontana, Robert. "A Brief History of UWB Communications." White paper, Multispectral Solutions.
8. Bennett, C. Leonard, and Gerald F. Ross. "Time-Domain Electromagnetics and Its Applications." *Proceedings of the IEEE* 66, no. 3 (March 1978).
9. Scholtz, R. A. "Multiple-Access with Time-Hopping Impulse Modulation." *Proc. 1993 MILCOM* (1993): 447–450.
10. Bass, Thomas A. "Think Tanked." *Wired Magazine* (December 1999).
11. First report and order in ET Docket No. 98-153, 17 FCC Rcd 7435 (2002), April 22, 2002.
12. Batra, Anuj, J. Balakrishnan, A. Dabak, R. Gharpurey, P. Fontaine, J. Lin, J.-M. Ho, and S. Lee. "Physical Layer Submission to 802.15 Task Group 3a: Time-Frequency Interleaved Orthogonal Frequency Division Multiplexing (TFI-OFDM)." IEEE P802.15-03/142r0, March 11, 2003.
13. Aiello, Roberto, T. Larsson, D. Meacham, Y. Kim, and H. Okado. "Multi-Band Performance Tradeoffs." IEEE 802.15-03/209r0, May 2003.
14. See www.multibandofdm.org, now www.wimedia.org.

2

UWB Spectrum and Regulations
by Robert Sutton

UWB is defined as available spectrum. As of today, it has been regulated in the United States, while the rest of the world is still working towards a global spectrum allocation. UWB rules are different from any other existing spectrum regulations, and UWB transmitters must meet very stringent requirements to be allowed to operate. This chapter describes the FCC rules required to test a UWB transmitter for compliance.

2.1 Regulatory Testing of UWB Devices

A UWB radio must comply with the administrative and technical requirements of both parts 2 and 15 of Title 47 of the Code of Federal Regulations (47 CFR) to be legally used, imported, or marketed in the United States. Part 2 covers the general rules and regulations regarding frequency allocations and radio treaty matters. Part 15 covers the radio frequency (RF) devices themselves. While both sections contain information needed to bring a UWB device to the marketplace, Part 15 contains the necessary technical requirements that are the subject of this chapter [1].

2.2 UWB Regulatory Terminology

Before we begin the regulatory discussion, a short introduction to some terminology is beneficial. The following is a brief glossary of some of the terms that will be used throughout this chapter [1–3].

- *EMC*. Electromagnetic compatibility. EMC requirements stipulate that a device shall not cause interference within itself or in other devices or be susceptible to interference from other devices.
- *FCC*. Federal Communications Commission. Established by the Communications Act of 1934, the FCC is the federal agency in charge of overseeing interstate telecommunications, as well as all communications services originating and terminating in the United States.

- *CISPR.* Comité International Spécial des Perturbations Radioélectriques (Special International Committee on Radio Interference). This body is concerned with developing norms for detecting, measuring, and comparing electromagnetic interference in electric devices. Some members are also in the International Electrotechnical Commission (IEC). It was founded in 1934.
- *ANSI.* American National Standards Institute. ANSI is a private, nonprofit organization that administers and coordinates the United States' voluntary standardization and conformity assessment system. It promotes and facilitates voluntary consensus standards and conformity assessment systems.
- *MPE.* Maximum permissible exposure. The MPE is the root mean square (RMS) and peak electric and magnetic field strengths, their squares, or the plane-wave equivalent power densities associated with these fields and the induced and contact currents to which a person may be exposed without harmful effects and with an acceptable safety factor.
- *47 CFR.* Title 47 of the Code of Federal Regulations. FCC rules and regulations are codified in the various parts (e.g., Part 15, Radio Frequency Devices) and subparts (e.g., Subpart F, Ultra-Wideband Operation) of 47 CFR. The rules are initially published in the *Federal Register*. The FCC does not maintain a database of its rules; nor does it print or stock copies of the rules and regulations. That task is performed by the Government Printing Office (GPO). After October 1 of each year, the GPO compiles all changes, additions, and deletions to the FCC rules and publishes an updated Code of Federal Regulations. The following definitions are taken directly from the sections of the 47 CFR regulatory text:
  — *UWB bandwidth.* For the purposes of this subpart, the UWB bandwidth is the frequency band bounded by the points that are 10 dB below the highest radiated emission as based on the complete transmission system, including the antenna. The upper boundary is designated f_H and the lower boundary is designated f_L. The frequency at which the highest radiated emission occurs is designated f_M; it must be contained within the UWB bandwidth.
  — *UWB transmitter.* An intentional radiator that, at any point, has a fractional bandwidth equal to or greater than 0.2 (20 percent if expressed as a percentage) or has a UWB bandwidth equal to or greater than 500 MHz, regardless of the fractional bandwidth.
  — *UWB communications system.* A system as defined in this section involving the transmission, emission, and/or reception of radio waves for specific UWB communications purposes.
  — *Center frequency.* The center frequency, f_C, equals $(f_H + f_L)/2$. If the UWB bandwidth is purely symmetrical around the highest radiated-emissions point, then f_M equals f_C.

- *Fractional bandwidth.* The fractional bandwidth equals $2(f_H - f_L)/(f_H + f_L)$ or alternatively $(\text{UWB BW})/f_C$. Both terms can also be represented as percentages by multiplying the result by 100.
- *EIRP.* Equivalent isotropic radiated power. The product of the power supplied to the antenna and the antenna gain in a given direction relative to an isotropic antenna.
- *Handheld.* As used in Subpart F, a handheld device is a portable device, such as a laptop computer or PDA, that is primarily handheld while being operated and does not employ a fixed infrastructure.
- *Digital device.* An unintentional radiator that generates and uses timing signals or pulses at a rate in excess of 9 kHz and uses digital techniques.
- *Intentional radiator.* A device that intentionally generates and emits radio-frequency energy by radiation or induction.
- *Unintentional radiator.* A device that intentionally generates radio-frequency energy for use within a device or that sends radio-frequency signals by conduction to associated equipment via connecting wiring, but which is not intended to emit radio-frequency energy by radiation or induction.
- *Class A digital device.* A digital device that is marketed for use in a commercial, industrial, or business environment, exclusive of a device which is marketed for use by the general public or is intended to be used in the home.
- *Class B digital device.* A digital device that is marketed for use in a residential environment, notwithstanding use in commercial, business, and industrial environments.

2.3 Testing Infrastructure

We begin our discussion of UWB regulatory testing with a description of the infrastructure necessary to support the measurements. In the case of UWB measurements, this refers to a calibrated measurement environment and a complementary set of instrumentation.

2.3.1 Anechoic Chambers

Test engineers use anechoic chambers to make radio measurements in a noise-free, weatherproof, and secure environment. Effectively eliminating the uncertainty associated with the ambient RF spectrum helps the overall accuracy and repeatability of the measurements. At the same time, the anechoic chamber must correlate well with the theoretical performance of a standard reference site. This ensures the veracity of the measurements. (See Figure 2.1.)

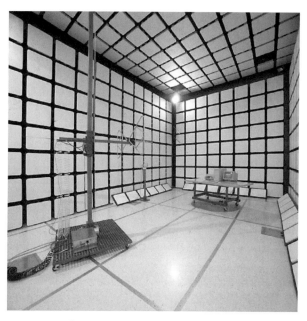

Figure 2.1 EMC semianechoic chamber.

Anechoic chambers comprise two basic parts, each with separate functions. The first part is a Faraday cage that prevents electromagnetic radiation from seeping into the measurement environment. This allows measurements to be made without the influence of the ambient RF spectrum. The amount of shielding effectiveness the cage provides depends on several factors: metallic construction material, material thickness, apertures, and seam continuity. The cage also serves another function: it contains the electromagnetic radiations inside the chamber. This helps to keep the external RF environment clean.

The second part of the anechoic chamber is the actual anechoic material itself. This material lines the surfaces of the cage and is used to attenuate the propagation of electromagnetic waves within the chamber. The layers are composed of ferrite tile material and foam absorber material. (See Figure 2.2.)

It is possible for the solution to implement only one of the two layers (this usually depends on the frequency range of interest). The measurement application determines the need for a combination of materials or the use of a single material. Most modern radio measurements are made in one of two types of chambers: semianechoic or fully anechoic.

The "semi" in *semianechoic* refers to the fact that a perfectly reflecting ground plane is implemented in the chamber. All other surfaces are covered with anechoic material. The ground plane serves as a reference plane approximating an open-area test site (OATS). The OATS, as defined in ANSI C63.7 (American National Standard Guide for Construction of Open-Area Test Sites for Performing Radiated Emission Measurements), is the required measurement reference for most EMC-

Figure 2.2 Radio frequency anechoic materials.

related radiated-emissions measurements [4]. Measurements for electromagnetic compatibility require a hybrid solution using a ferrite and matched foam combination to cover a wide range of measurements. Figure 2.3 is an EMC chamber that accommodates measurements from 30 MHz to 40 GHz in one facility.

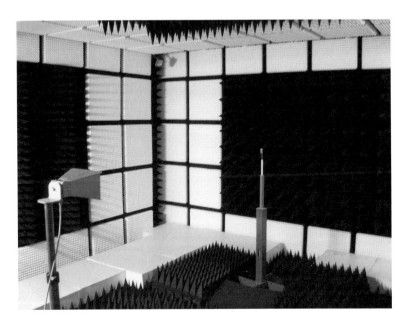

Figure 2.3 Fully anechoic chamber.

The term *fully anechoic* refers to the fact that all surfaces of the chamber are covered with anechoic material.

A free-space test environment, qualified per IEEE Standard 149-1979 (IEEE Standard Test Procedures for Antennas) [5], is suitable for most EMC-related free-space radiated-emissions measurements. For example, a completely anechoic chamber suitable for UWB antenna measurements would require only a foam-based absorber solution since the measurements occur at frequencies higher than 960 MHz.

2.3.2 Alternative Measurement Environments

Alternative environments can be used to make UWB radiated-emissions measurements. As discussed in Section 2.3.1, the preferred reference for radiated-emissions measurements is the OATS. What the OATS lacks in sophistication it makes up for in economies of scale. It is by far the lowest-cost facility to implement; however, with today's crowded and disruptive radio spectrum, susceptibility to weather, and lack of security, fewer and fewer facilities are using the OATS. Even covered OATS (sites with radio-transparent enclosures to provide weatherproofing and security) are having to be reevaluated for their effectiveness. As higher and higher frequency measurements are demanded, the radio transparency of the enclosure material becomes suspect. In addition, trapped particles and dirt on the enclosure body itself can influence the facility's propagation statistics.

Another alternative environment for radiated-emissions measurements would be the reverberation chamber (also known as the mode-tuned or mode-stirred chamber) (Figure 2.4). Such solutions generally use a mechanical tuner or stirrer to change the boundary conditions in the chamber and, thus, to change the structure of the electromagnetic fields in the test environment. Radiated-emissions testing in a reverberation chamber is considered a stochastic process whereby the mechanical tuner "stirs" the different "modes" of the enclosure to provide a statistical field uniformity within the test volume. The chamber does not contain any anechoic materials as this would dampen the electric fields and prevent the chamber from reverberating. The size of the chamber dictates the frequency range of operation. To operate at the lower frequencies required for EMC measurements, the chamber size must be significant. This is one reason reverberation chambers have not become popular for emissions measurements below 200 MHz. Additionally, polarization data cannot be measured with a reverberation chamber since it depends on a scattered field to generate its uniformity. The reverberation chamber is a useful environment for making low-level measurements. Due to the mode stirring, it takes a little longer to get the results.

Finally, another option is to use high-frequency transverse electromagnetic (TEM) cells. Physically, these devices are expanded coaxial waveguides. The device under test is placed between the outer conductor and the septum. The electric fields generated by the device under test create a voltage drop between the septum

Figure 2.4 EMC reverberation chamber.

and the outer conductor. This voltage can then be measured by a receiver or spectrum analyzer.

To make a reasonable correlation between one of these cells and an OATS, the engineer must use a sophisticated measurement algorithm, which requires that the equipment under test be rotated to expose each side (or combination of sides) to the cell's septum. This must be done because the septum is stationary, so the equipment under test has to move with respect to it in order to capture all of the energy. Although TEM-based measurement systems are handy and compact, they do have some limitations. The dual-port classical TEM cells are limited with respect to their frequency coverage because their size is inversely proportional to the frequency range. In order to implement high-frequency measurements, the TEM cell would have to be very small. A variety of single-port tapered TEM cells extend the frequency range of operation without compromising the cell size. These cells are constructed with broadband absorber terminations and resistive loads, as well as offset septums. That is, the field-generating or -receiving conductor is asymmetrically placed with respect to the outer conductor. These cells are limited in frequency by their absorber and resistive termination quality and the nonuniformity of the electric field distribution between the septum and the outer conductor. Finally, all varieties of TEM cell solutions are very susceptible to cable placement and manipulation. Cables (e.g., power-line, communication) can present resonant structures within the cell's cavity and, as such, provide for additional measurement uncertainties that are difficult to separate from the equipment's intended emissions profile.

Each one of these alternates has its own advantages and disadvantages. If any measurements are to be performed in alternative environments, special attention should be paid to the frequency limitations and calibration methods in order to obtain meaningful data. Furthermore, many regulatory agencies may not recognize measurements made in alternative environments. In these instances, the engineer will be forced to use the data for preliminary understanding of the radio's behavior only and not as a regulatory grade measurement.

2.3.3 Wideband or Narrowband

Two functional perspectives, system design and regulatory evaluation, can be used to differentiate UWB and narrower-band radio systems. As radio architects, many engineers are familiar with the general differences between UWB and narrowband system design concepts. Such aspects as the impulse response of the radio's front end, sophisticated fading models for propagations analysis, and the inclusion of energy-capture terms in the link budget can improve prediction and performance analysis; however, this chapter is concerned with the regulatory aspects of UWB communications systems and the distinction between wideband and narrowband in that context.

From a regulatory perspective, the differences between classical narrowband and UWB systems reside in the measurement details of the test instrumentation. Whether a system is determined to be wideband or narrowband from a regulatory standpoint is actually determined by the measurement receiver's resolution bandwidth or impulse response. Whether a measurement system is affected or influenced by a wideband or narrowband signal from a regulatory standpoint is determined by the support peripherals. The fact that the mid-band UWB measurements are made with a receiver resolution bandwidth of 1 MHz means that the signals are certainly considered wideband. Furthermore, it means that the regulatory engineer's focus should be on the accuracy of the measurement. Measuring wideband signals in the presence of noise presents multiple challenges that are usually not well posed and require the test engineer to monitor the measurement system more closely. Directly stated, this means that the limits of the measurement system should be well understood. Such items as receiver attenuation levels and mixer input limits, preselector options and cutoff points, low-pass and high-pass filter banks, and preamplifier noise figures and saturation points should all be well characterized to avoid inaccurate measurements.

2.3.4 Test Instrumentation

The type of test instrumentation used for UWB regulatory work is directly related to whether the spectral measurements are made in the time or the frequency domain. There are advantages to each methodology tied directly to what the engineer would like to achieve and the measurement environment he or she will be operating in.

2.3.5 Time Domain

Measurements in the time domain require test environments and instrumentation with large instantaneous bandwidths. This is due to the fact that time-domain measurements are theoretically full bandwidth measurements [6]. (See Figure 2.5.)

The measurement system comprises a series of wideband transducers, digital sampling heads, a digital oscilloscope, and positioning equipment. In addition, preamplifiers may also be needed to increase the dynamic range of the system. A lot of postprocessing of the data is involved as well. The regulatory limits are provided in the frequency domain, whereas the measurements are captured in the time domain. This means that the entire dataset must be converted to the frequency domain via a Fourier transformation, usually accomplished by an FFT algorithm.

Further processing may be required to account for instrument bandwidth limitations, the transducer factors, the gain of the preamplifiers, cable losses, detection weighting, and so forth. The process is not an easy one and care must be applied with the setup and processing of the transformation so that additional spectral components are not produced through transformation or weighting errors.

In addition, the data must be spatially maximized, just as it would be in pure frequency-domain measurements. That is, the equipment under test must be rotated through 360° in the azimuth, and the receive transducer must be scanned in height from 1m to 4m (provided that the test environment is semianechoic). All of this is for the standard 3m range length. Accordingly, to achieve this, the test environment must be carefully assessed for pulsed performance.

Figure 2.5 Time-domain measurement in a fully anechoic chamber.

When these measurements are done in a semianechoic or fully anechoic chamber, the absorber performance is critical. Additional reflections from the specular regions on the side walls, ceiling, and end walls may distort the received pulse, compromising the accuracy of the received measurements. An understanding of the room's performance through pulsed characterization measurements is very helpful in diagnosing any potential problems.

In summary, time-domain measurements can be used to some degree of success for regulatory measurements; however, doing so requires a thorough understanding of all facets of the measurement chain. In addition, limitations within the measurement itself require that special care be taken so as not to introduce additional measurement errors through inadequate dynamic range, transformation errors, or the application of weighting algorithms to simulate detector responses.

2.3.6 Frequency Domain

Measurements performed in the frequency domain require flexible measurement environments and can rely on instrumentation with limited bandwidths as long as the measurements are staged properly. This is due to the fact that frequency-domain measurements are band limited in nature; that is, the intermediate frequency (IF) bandwidth of the measurement instrument is significantly smaller than the bandwidth of the radiated waveform [6].

Figure 2.6 UWB EMC measurement system.

The frequency-domain measurement system comprises a series of band-specific antennas, receiver and spectrum analyzers, preamplifiers, filters, and positioning equipment. The data postprocessing necessary for a frequency-domain measurement is much less than that required for a time-domain measurement. To begin with, the regulatory limits are provided in the frequency domain, the same domain in which the measurements are captured. This means that the entire dataset can be operated on directly without the need for a domain transformation.

Postprocessing must be done to account for the antenna factors, the gain of the preamplifiers, cable losses, filter losses, and so forth. The detection types and resolution, as well as the video bandwidths, of the measured data are all accounted for in the spectrum analyzer's data, so no special weighting is required to account for these. (See Figures 2.6 and 2.7.)

Most measurement instrumentation has resident firmware that compensates for the raw measurement data in real time to take into account antenna factors, external preamplifier gains, cable losses, and so forth. This ability can be an advantage for manual measurements, allowing the composite data to be instantly displayed on the instrument's screen and compared to a limit line. It may not be so convenient if the raw data is required. Access to the raw spectrum data is useful as it can be postprocessed with different transducers to determine different spectrum responses. The most flexible solution is to use an EMC software package with UWB measurement utilities that can produce the raw and compensated data, along with all of the factors used during the measurement.

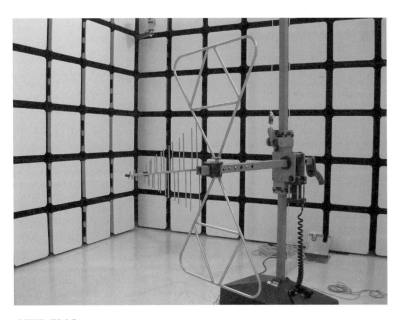

Figure 2.7 UWB EMC measurement antenna.

Like time-domain measurements, the frequency-domain data must be spatially maximized. That is, for a 3m certified measurement, the equipment under test must be rotated through 360° in the azimuth, and the receive antenna must be scanned in height from 1m to 4m. Accordingly, to achieve this, the test environment must be carefully assessed for swept frequency performance and be in compliance with the normalized site attenuation specified in ANSI C63.4 (American National Standard for Methods of Measurement of Radio-Noise Emissions from Low-Voltage Electrical and Electronic Equipment in the Range of 9 kHz to 40 GHz) [7]. To ensure satisfactory performance in a semianechoic chamber, a quality factor related to the normalized site attenuation (NSA) is assigned to the test area or quiet zone. For measurements made in a fully anechoic test environment, the quality factor is measured using a free-space transmission loss (FSTL) measurement over the quiet zone. These quality factors specify a tolerable amount of electric-field variation over the quiet zone, taking into account the measurements that will be undertaken in the test environment.

In summary, the frequency domain is well suited for making regulatory measurements to existing EMC standards. In fact, since UWB regulatory limits are specified in the frequency domain, the errors that could occur from any domain transformations are eliminated.

2.4 Regulatory Overview

More than likely, the regulatory filing for a UWB radio will be a composite filing, comprising the requirements of several sections of Part 15. Depending on the product's configuration, the following sections of Part 15 will be utilized [1, 8].

2.4.1 Subpart A—General

This subpart sets out the regulations under which an intentional, unintentional, or incidental radiator may be operated without an individual license. It comprises many sections that contain the technical specifications, administrative requirements, and other conditions relating to the marketing of Part 15 devices. Although all of these sections are general in nature and, therefore, have some bearing on a UWB regulatory submittal, several are more important than others. A brief review of the salient points follows.

- Measurement standards—§ 15.31. This area of Subpart A provides an overview of the various field-strength measurements, low-frequency considerations, measurement topologies, and more. It also details where to get more information on the measurement procedures that the FCC relies on to determine compliance with specific technical requirements. Information on configuring a measurement for intentional and unintentional radiators, such as UWB devices, is contained in ANSI C63.4. ANSI has published this document over the years to add improvements and harmonize it with global regulatory standards. The regulatory engineer should reference § 15.31 for advice on which version of ANSI C63.4 is the latest and appropriate for measurement use.

- Frequency range of radiated measurements—§ 15.33. Obtaining the proper upper and lower limits of radiated-emissions measurements for a UWB device requires that it be looked at from two different perspectives.

    —As an intentional radiator the frequency limits are based on factors related to f_C (center frequency).

    —As an unintentional radiator the limits are determined by the highest frequency generated or used by the device or on which the device operates or tunes.

- Measurement detector functions and bandwidths—§ 15.35. The FCC prescribes different measurement detectors and bandwidths for different measurement types and frequency ranges. Table 2.1 shows a combination of resolution bandwidths and detector types used for measurements compliant with subparts B, C, and F.

The following are some items to note:

- Several types of detectors are used for electromagnetic-compatibility measurements [9]. Included are two methods for detecting the average response of a signal. One type of measurement relies on power averaging to get the RMS-detected values of a signal. The other type of measurement relies on voltage averaging to provide the electromagnetic interference average (AVG).
- The peak (PK) detector is used to report the maximum value of the signal encountered in each measurement bin.
- The quasipeak (QP) detector is a weighted form of peak detection. The measured value of the QP detector drops as the repetition rate of the measured signal decreases. The QP detector is a way of measuring and quantifying the annoyance factor of a signal.

	Frequency (MHz)	RBW	Detector
Subparts B and C	0.009–0.150	200 Hz	QP
	0.150–30	9 kHz	QP
	30–1,000	120 kHz	QP
	> 1,000	1 MHz	AVG
Subpart F	0.009–0.150	200 Hz	QP
	0.150–30	9 kHz	QP
	30–960	120 kHz	QP
	> 960	1 MHz	RMS

Table 2.1 Resolution Bandwidth and Detector Types for EMC Testing

- The AVG detector uses voltage averaging by averaging the linear voltage data of the envelope signal measured during the bin interval. It is often used in electromagnetic-interference testing to measure narrowband signals that might be masked by broadband impulsive noise.
- The RMS detector works by power averaging. Measurements are made by taking the square root of the sum of the squares of the voltage data measured during the bin interval, divided by the characteristic input impedance of the spectrum analyzer, normally 50Ω. This type of power averaging calculates the true average power of the signal being detected and is the best way to measure the power of complex signals.
- Most preliminary measurements are performed using the PK detector for speed and try to capture the worst case emissions. Then, the peak emissions are identified, and the final measurements are performed using the required detector.
- ac power-line conducted-emissions limits in subparts B and C also have limits based on AVG detection.
- For subparts B and C measurements above 1,000 MHz, use an AVG-type detector, provided the peak emission is not more than 20 dB above the limit.

2.4.2 Subpart B—Unintentional Radiators

This subpart defines the regulations under which an unintentional radiator is authorized and includes the limits for unintentionally radiated emissions for digital circuitry. Few of the sections in Subpart B impact UWB regulatory issues. A brief review of these salient points follows.

- Equipment authorization of unintentional radiators—§ 15.101. Part (b) of this section states that receivers operating above 960 MHz or below 30 MHz, with the exception of CB radios and radar detectors, are exempt from having to comply with the technical provisions of Subpart B.
- Radiated emission limits—§ 15.109. The limit lines for unintentionally radiated emissions due to digital circuitry are described.

2.4.3 Subpart C—Intentional Radiators

This subpart defines the regulations under which an intentional radiator is authorized and includes the limits for a number of intentionally radiated emissions; however, the details for the intentionally radiated limits for UWB come in Subpart F. Subpart C is referenced in Subpart F to use the limits of intentional radiation due to unwanted emissions not associated with fundamental UWB frequencies and to use ac conducted-line emission limits.

- Conducted limits—§ 15.207. The limits of the RF voltages that are conducted back onto the public-utility power lines are prescribed.

- Radiated emission limits—§ 15.209. The limits in this section are based on the frequencies of the unwanted emissions and not the fundamental frequency.

2.4.4 Subpart F—UWB Operation

The scope of this particular subpart deals with the authorization of unlicensed UWB transmission systems. It forms the core regulatory requirements needed to bring a UWB device to market. Since this book is about UWB communications systems, discussions concerning UWB imaging and vehicular systems are not included in this chapter. The following will be combined into one section and covered in more detail in this chapter:

- Technical requirements for indoor UWB systems—§ 15.517
- Technical requirements for handheld UWB systems—§ 15.519
- Technical requirements applicable to all UWB systems—§ 15.521

2.4.5 Basic UWB Radio Overview

The regulations that apply to a UWB communications device can be confusing. To understand them, referring to a generic block diagram helps as it is easier to analyze the radio from the point of view of a regulator.

Figure 2.8 provides a very basic overview of a pulse-modulated RF radio topology [10]. The UWB implementation scheme is very important because it will determine how the regulators decide which types of tests or exercise modes the radio requires. Several of the same techniques can be used to identify the various radio subsections

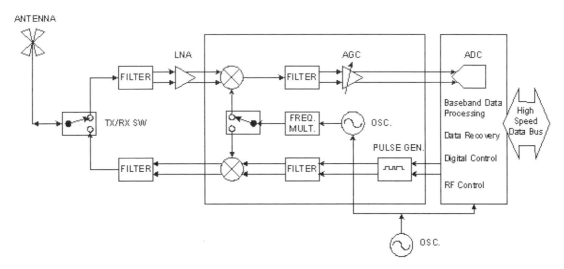

Figure 2.8 Generic UWB radio topology.

regardless of whether the radio is using baseband-pulse, pulse-modulated RF or a different scheme to generate a UWB signal.

With regard to regulatory work, we are interested in three main areas of the UWB device: the transmitter, the receiver, and the control circuitry.

The transmitter is the heart of the system and is the part of the device most people refer to when they speak generically of the radio. It encompasses the generator and shaping circuitry and arguably any additional output filtering and baluns required to produce the communications waveform. The type of UWB signal generation and modulation used are defined within the transmitter and, as such, will dictate what types of test modes must be exercised. It is worth mentioning that the antenna is part of the transmission system; however, it has multiple roles and will be discussed separately.

The receiver section of the radio is often overlooked because of its relative silent state with respect to the transmitter. Nonetheless, the regulations require that any unintentional radiations emitted by the receiver circuitry be investigated with respect to the proper limits.

Digital control circuitry is handled very specifically in the case of UWB radio. Transmitter control, bus control, clock sequences, and other digital signaling are all limited differently depending on where their emissions show up in the spectrum. Several different rules must be applied to determine the correct limit lines to apply to these emissions. A detailed discussion of the rules devoted to technical requirements for indoor and handheld UWB systems is carried out in Section 2.4.6.

We are also interested in three secondary, but very important, support items: antennas, peripherals, and external power supplies. These items are very important and are accounted for in the measurement setup and the application of the correct subpart.

UWB regulatory matters require that the antenna be a nonseparable integrated part of the radio. Furthermore, only the antenna for which the UWB radio is authorized may be used with the device [1, 2]. At the moment, this means that separate licenses must be garnered for each device configuration using alternate antennas until a multiple antenna listing scheme is implemented. Due to the complicated interaction and dependence of the radiated waveform on the antenna, it is regarded as a key part of the design and regulation of the radio. The antenna to be used with a particular UWB product must be carefully selected to ensure optimum operation. Such factors as pattern coverage, pulse fidelity, return loss, gain, and efficiency should all be considered [11]. From a regulatory standpoint, the antenna influences the UWB bandwidth and the spectral shaping; both parameters are measured against limits published in the standards.

Peripherals are another matter that must be dealt with in making regulatory measurements (Figure 2.9). The radio is considered to be the main device under test; however, any item that interfaces with it is considered a support peripheral. Many UWB radios incorporate local area network (LAN), USB, serial, and other interfaces. First-generation radios may contain cabled connections between the radio and

Figure 2.9 UWB radio and peripherals.

these peripherals. These interfaces are used to transfer data between the radio and external equipment, such as computers, video-distribution units, and other items. In these circumstances, the radio should be configured as an integrated system with all of these ports connected and exercised with the appropriate peripheral devices. This has to be done in order to determine the effect on the radiated- and conducted-emissions spectrum. Usually, it is only required that one type of port on devices that exhibit multiple ports of the same variety be exercised. The additional ports, however, must still be terminated with a representative load.

External power supplies are considered to be special-case peripherals. Different types of power-supply topologies can yield surprisingly different measurement results. Most unfiltered power supplies will couple radio energy into the emissions spectrum. This can be problematic if the emissions exceed the relative limits. The phenomenon can be further exacerbated by the load constraints put on the supply and the temperature of operation. The majority of power supplies convert ac power to switched-mode dc power and will radiate broadband low-frequency radio noise as well as conduct ac power-line noise.

Peripherals and their cabling must be included in the measurements for several reasons. Oftentimes, radio frequencies couple to the external ports and radiate or conduct through the cabling. In these cases, the equipment under test becomes an extended system, with the radio as the centerpiece and the peripheral equipment, support cabling, and power supply completing the ensemble. ANSI C63.4 provides guidelines for equipment spacing, references for test distance points, and cable routing for tabletop subsystems.

Table 2.2 specifies which part of the UWB device has to comply with which part of the regulation.

47 CFR, Part 15, Subpart	Application
A	All parts of the UWB device (measurement standards, frequency range of measurements, detector functions and bandwidths are discussed)
B	Digital circuitry that is not related to the operation or control of the transmitter
C	Digital circuitry that is used to control the transmitter but that is not intended to be radiated though the antenna, AC conducted mains
F	UWB radio transmitter, digital circuitry that has spectrum radiated though the antenna

Table 2.2 Subpart Application Table for UWB Measurements

A UWB wireless transceiver would contain all basic parts of a UWB radio and would therefore need to comply with all subparts listed in the table. Another example would be a simple UWB transmitter used as a test source for an evaluation kit. In this case, there may be no testing to Subpart B, only to subparts C and F, with guidance from Subpart A. In some products or measurement scenarios, it may be difficult to differentiate the source of a radiated emission clearly. Such a case could arise when common control circuitry is used to handle both the transmit and the receive portion of a radio. Details about the determination and differentiation of various signal types are complicated and beyond the scope of this chapter.

2.4.6 Technical Requirements for Indoor and Handheld UWB Systems

The scope of this particular section deals with unlicensed UWB transmission systems used for communications services. It follows the technical requirements for UWB communications systems outlined in the 47 CFR, Part 15, Subpart F [1, 8]. More specifically, the sections dealing with communications systems, § 15.517, § 15.519, and § 15.521, will be covered. Other than a brief mention, discussions concerning UWB imaging and vehicular systems are not included in this section.

Items tested to CFR 47, Subpart F, § 15.517 and § 15.519, follow the same measurement regimen with some slight differences.

2.4.6.1 General Requirements

As the name implies, indoor UWB systems are restricted to indoor operation only and, as such, must comply with those rules. All indoor UWB communications applications (audio-visual applications, WUSB, various types of home networking, and so forth) fall under this category and must comply with the following general guidelines.

1. Proof of indoor operation: Usually reliance on an ac power connection is sufficient.

2. Controlled radiation pattern: The emissions from the UWB equipment shall not be intentionally directed outside of the building in which the equipment is located.
3. Outdoor antennas: The use of any outdoor radiating devices is prohibited.
4. Field disturbance sensors: Communication systems located inside metallic or underground storage tanks are considered indoor systems provided the emissions are directed towards the ground.
5. Limited transmission: A communication system shall transmit only when the intentional radiator is sending information to an associated receiver. Continuous transmission is prohibited.
6. Labeling requirements: UWB systems to be operated indoors must have the following instructions (prominently displayed on the device or in the user's manual): "This equipment may only be operated indoors. Operation outdoors is in violation of 47 U.S.C. 301 and could subject the operator to serious legal penalties."

Handheld devices may be used for outdoor or indoor operation. These UWB devices must be handheld as per the previous definition and must comply with the following general guidelines:

1. Transmission acknowledgement: The device shall transmit only when the intentional radiator is sending information to an associated receiver. There is a 10 second transmission limit (without an acknowledgement) on UWB transmission under this provision. An acknowledgement of reception must continue to be received by the transmitter every 10 seconds, or the device must stop transmitting.
2. Outdoor antennas: The use of antennas mounted on outdoor structures is prohibited. Antennas may be mounted only on handheld UWB devices.

2.4.6.2 Frequency Range to Consider in Compliance Measurements

To obtain the proper upper and lower limits of radiated-emissions measurements for a UWB device requires that we look at it from two different perspectives: as an intentional radiator and an unintentional radiator. We have described all frequency bands applicable for investigation but have limited our spectrum mask example in the figures to a lower frequency of 30 MHz and an upper frequency of 40 GHz for practicality. In addition, keep in mind that this section only determines the minimum and maximum frequency range of the measurements to be taken and not the amplitude limits to which they are compared. Particular limits are described in each of the remaining sections of this chapter (via tables and figures).

Intentional radiator: The UWB device in its transmit mode behaves as an intentional radiator. The highest frequency used to determine the frequency range over which the radiated measurements are made will be based on the center frequency, f_C;

DUT Center Frequency (f_c)	Upper Frequency Limit of Measurement
< 10 GHz	The lesser of $10 \times f_c$ or 40 GHz
10 GHz ≤ f_c < 30 GHz	The lesser of $5 \times f_c$ or 100 GHz
≥ 30 GHz	The lesser of $5 \times f_c$ or 200 GHz

Table 2.3 Upper Frequency Limit for UWB-related Intentionally Radiated Emissions

however, if a higher frequency is generated within the UWB device, then this higher frequency shall be used for the calculation of the upper measurement limit.

An intentional radiator's spectrum must be investigated from the lowest RF signal generated (down to 9 kHz) up to the maximum frequency shown in Table 2.3 as derived from § 15.33(a), or the maximum frequency determined from the calculation, f_C + 3/(pulse width in seconds), whichever is greater.

Note: If the intentional radiator contains a digital device, regardless of this digital device's function (controlling the intentional radiator or additional unrelated functions and controls), the frequency range of investigation shall be according to Table 2.3 or the range applicable to the digital device, as shown in Table 2.4; whichever results in the higher frequency range being measured.

Unintentional Radiator: The UWB device produces spectra due to digital devices in the control circuitry. From this perspective, it behaves as an unintentional radiator. The frequency used to determine the range over which the radiated measurements are made will be based on the highest fundamental frequency of these emissions.

An unintentional radiator's spectrum must be investigated from the lowest RF signal generated (down to 9 kHz) up to the maximum frequency shown in Table 2.4, as derived from § 15.33(b).

Highest frequency generated or used in the device or on which the device operates or tunes	Upper Frequency Limit of Measurement (MHz)
< 1.705 MHz	30 MHz
1.705 MHz ≤ f < 108 MHz	1,000 MHz
108 MHz ≤ f < 500 MHz	2,000 MHz
500 MHz ≤ f ≤ 1,000 MHz	5,000 MHz
> 1,000 MHz	The lesser of $5 \times f$ or 40 GHz

Table 2.4 Upper Frequency Limit for UWB-related Unintentionally Radiated Emissions

Figure 2.10 Dual polarized measurement horn for UWB measurements.

2.4.6.3 UWB Bandwidth Restrictions

According to § 15.517 and § 15.519 of Subpart F, the bandwidth (−10 dB) of UWB communications devices must be contained within the 7.5 GHz of spectrum contained between 3,100 and 10,600 MHz. The frequency at which the highest radiated emission occurs, f_M, must also be contained within the UWB bandwidth of the product.

The UWB bandwidth should be determined by maximizing the signal from the radio under test. This requires taking into account both horizontal and vertical measurement polarizations, full azimuth scans, and height scans over the ground plane. If tests are being performed in a fully anechoic chamber environment, then the height scan is not necessary.

Since the UWB bandwidth must be determined from the composite maximized signal, the correct tools are necessary to provide accurate measurement results. A fixed aperture antenna, such as a dual polarized horn that captures both polarizations simultaneously and covers the prescribed frequency bandwidth, is a time-saving, practical transducer to employ for these measurements [6, 12]. (See Figure 2.10.)

2.4.6.4 UWB-Specific Radiated-Emissions Limits up to 960 MHz

The general emissions limits applicable to UWB communications systems are derived in part from § 15.209 of Subpart C. Table 2.5 provides the limits for emissions below 960 MHz.

38 Chapter 2

Frequency (MHz)	Measurement Distance (m)	Emissions Limit (μV/m)	Emissions Limit (dBμV/m)
0.009–0.490	300	2,400/F(kHz)	67.6–20Log[F(kHz)]
0.490–1.705	30	24,000/F(kHz)	87.6–20Log[F(kHz)]
1.705–30	30	30	29.5
30–88	3	100	40.0
88–216	3	150	43.5
216–960	3	200	46.0

Table 2.5 Emissions Limits Up to 960 MHz Applicable to UWB Communications Systems

UWB devices employ high internal frequencies. For this reason, tests down to frequencies below 30 MHz are quite rare. The standard has provisions for testing down to 9 kHz; however, for most commercial UWB communications, a 30 MHz lower boundary is sufficient. This is shown in Figure 2.11.

Several items must be noted for measurements executed in this frequency range. Careful attention must be paid when using the measurement table. If it is necessary to investigate lower frequencies (9 kHz to 30 MHz), you will see that the measurement range increases significantly. Using a smaller range to qualify a product at these frequencies would likely increase the measurement uncertainty due to near field errors. Electromagnetic scaling would have to be used carefully if appropriate. Any radio noise occurring at the emission limit band edges will have to comply with the stricter limits (e.g., an emission occurring at 88 MHz will have to meet the 40 dBμV/m limit rather than the 43.5 dBμV/m limit). The limit line is based on the QP-detected signal measured with a resolution bandwidth of 120 kHz. The video bandwidth and measurement time are set automatically by the receiver to comply

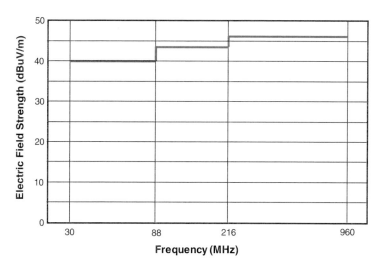

Figure 2.11 Emissions limits up to 960 MHz applicable to UWB communications systems.

UWB Spectrum and Regulations 39

Figure 2.12 EMC low frequency UWB measurements.

with CISPR publication 16 requirements. The limit line is expressed in dBμV/m (unit of logarithmic field strength referenced to microvolts) since that is the more common notation for global standardization of emission limits, and it is easier to work with logarithmic correction factors when postprocessing the data. Table 2.5 contains the limits expressed in both μV/m and dBμV/m for easy cross-referencing to the FCC subpart of concern. (See Figure 2.12.)

2.4.6.5 UWB-Specific Radiated-Emissions Limits Greater Than 960 MHz

The general emissions limits applicable to UWB communications systems are derived from Subpart F, § 15.517 and § 15.519. Refer to Table 2.6 for emissions limits above 960 MHz.

Frequency (MHz)	Measurement Distance (m)	Emissions Limit (dBm/MHz)	
		Indoor	Hand Held
960–1610	3	−75.3	−75.3
1610–1990	3	−53.3	−63.3
1990–3100	3	−51.3	−61.3
3100–10,600	3	−41.3	−41.3
> 10,600	3	−51.3	−61.3

Table 2.6 Emissions Limits Greater than 960 MHz Applicable to UWB Communications Systems

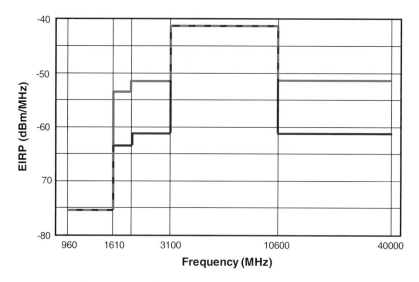

Figure 2.13 Emissions limits greater than 960 MHz applicable to UWB communications systems.

Figure 2.13 shows both the indoor and the handheld spectrum limits. This range covers the main band(s) of the UWB system and, therefore, contains the frequencies at which the highest radiated emissions should occur. The limit line is based on the RMS-detected signal measured with a resolution bandwidth of 1 MHz, a video bandwidth of at least 1 MHz, and a measurement (averaging time in the case of

Figure 2.14 EMC UWB high frequency measurements.

UWB Spectrum and Regulations 41

RMS detectors) time of 1 msec or less per frequency point. As stated in Section 2.4.6.4, any radio noise occurring at the emission limit band edges will have to comply with the stricter limits (e.g., an emission occurring at 1,610 MHz will have to meet the −75.3 dBm/MHz limit and not the −53.3 dBm/MHz indoor UWB limit).

All UWB emissions measurements performed at frequencies greater than 960 MHz are expressed in terms of their EIRP. The data and the limit line are expressed in units of dBm/MHz (unit of logarithmic power referenced to a milliwatt over a 1 MHz resolution bandwidth). (See Figure 2.14.)

Physically speaking, EIRP is the product of the power supplied to the antenna and the antenna gain in a given direction relative to an isotropic antenna [12]. Thus, for a given 3m measurement range and 1 MHz resolution bandwidth, we can relate the EIRP to the electric field strength represented in dBµV/m as

$$EIRP \text{ (dBm/MHz)} = E \text{ (dBµV/m)} - 95.3$$

Table 2.6 contains the limits for the indoor and the handheld spectrum masks, both expressed in terms of their EIRP values in dBm/MHz.

2.4.6.6 Intentionally Radiated–Emissions Limits for Digital Circuitry up to 40 GHz

Digital radiated emissions related to intentional radiators must be properly categorized in order to apply the correct regulatory limits. Two cases must be considered.

In the first case, digital circuitry that is used only to enable the operation of a transmitter and that does not control additional functions or capabilities must comply with the same limits as the UWB transmitter from Subpart F, § 15.517(c) and § 15.519(c); see Tables 2.5 and 2.6 and Figures 2.11 and 2.13. This is because the emissions may couple to the main antenna port and mix with the UWB waveform. If these emissions are present at the antenna terminal, they are not classified as coming from a digital device from a regulatory standpoint.

Frequency (MHz)	Measurement Distance (m)	Emissions Limit (µV/m)	Emissions Limit (dBµV/m)
0.009–0.490	300	2,400/F(kHz)	67.6−20Log[F(kHz)]
0.490–1.705	30	24,000/F(kHz)	87.6−20Log[F(kHz)]
1.705–30	30	30	29.5
30–88	3	100	40.0
88–216	3	150	43.5
216–960	3	200	46.0
960–40000	3	500	54.0

Table 2.7 Emissions Limits Applicable to Intentional Radiated Emissions from UWB Communications Systems Digital Circuitry

42 Chapter 2

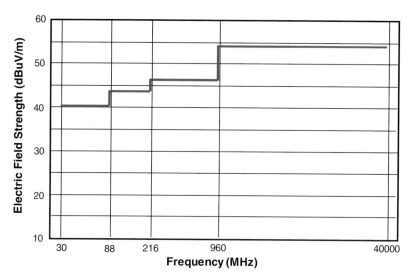

Figure 2.15 Emissions limits applicable to intentional radiated emissions from UWB communications systems digital circuitry.

In the second case, the digital circuitry emissions can be reclassified if it can be clearly shown that the digital emissions from the UWB transmitter are not intended to be radiated from the transmitter's antenna. In this situation, the emissions limits applicable are derived from § 15.209 in Subpart C; refer to Table 2.7 and Figure 2.15.

As with previous sections dealing with measurements over a broad frequency range, much attention to detail is required at the band edges, measurement distances, and frequency range of the measurements. A quick overview reveals that the limit line is based on the QP-detected signal measured with a resolution bandwidth of 120 kHz from 30 to 1,000 MHz. Above 1,000 MHz, the limit line is based on the average detected signal measured with a resolution bandwidth of 1 MHz, provided the peak emission is not more than 20 dB above the limit. The video bandwidths and measurement times are set automatically by the receiver to comply with CISPR publication 16 requirements [2].

2.4.6.7 Classification of Intentional Emissions

There are prescribed ways to determine whether the digital circuitry radiated emissions from intentional emissions are to be judged against the UWB-specific limits or against the § 15.209 limits. A detailed discussion of these techniques is beyond the scope of this chapter.

2.4.6.8 Unintentionally Radiated–Emissions Limits for Digital Circuitry up to 40 GHz

Digital circuitry that is not used, that is not related to the operation of a transmitter, or that does not control additional functions or capabilities related to the transmitter

UWB Spectrum and Regulations   43

Frequency (MHz)	Measurement Distance (m)	Emissions Limit (μV/m)	Emissions Limit (dBμV/m)
30–88	3	100	40.0
88–216	3	150	43.5
216–960	3	200	46.0
960–40,000	3	500	54.0

Table 2.8 Emissions Limits Applicable to Unintentional Radiated Emissions from UWB Communications Systems Digital Circuitry

is classified as an unintentional radiator from a regulatory standpoint. In this case, the device's emissions must comply with the Class B limits in § 15.109 of Subpart B. Refer to Table 2.8 and Figure 2.16.

For unintentionally radiated measurements that cover a broad frequency range, the same careful rationale is used as was used with the intentional radiators. There are, however, two main differences (aside from the fact that one signal is intentional and the other is unintentional) in the way the emissions are regulated.

One concerns the frequency range of regulatory application. The determination of the frequency range over which the measurements must be made, particularly how the upper limit is calculated, is different for the two emission types. Recall the discussion in Section 2.4.6.2.

The other difference concerns the categorization of the circuitry as a Class B digital device. This assumes that the devices will be used in a residential environment, as well as a commercial, business, or industrial environment. Using this categorization will guarantee that the limit lines are the most stringent.

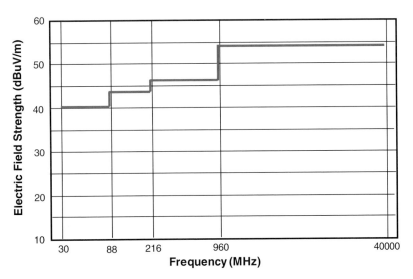

Figure 2.16 Emissions limits applicable to intentionally radiated emissions from UWB communications systems digital circuitry.

The same measurement bandwidths, detectors, and measurement rules are used to monitor the spectrum for unintentionally radiated emissions as are used for the collection of intentionally radiated data in Subpart C.

Note: If the device can be limited to Class A usage, that is, operation in a nonresidential environment (commercial, industrial, or business), then different limits must be applied to the device for unintentional emissions.

2.4.6.9 Identifying Unintentional Emissions

As with other emission types, practical methods can be used to identify unintentionally radiated emissions. This will be necessary because the resulting measured spectrum will yield a mix of unintentional and intentional emissions. Identifying whether a signal is from unintentional or intentional control circuitry requires knowledge of the various profiles of the specific control circuitry, as well as specific signal-identification techniques. In some cases, common control circuitry is used to handle both the transmit and the receive portions of a radio. The latter situation usually presents a more challenging measurement best left to a qualified laboratory to handle.

2.4.6.10 Spectral Line Emissions Limits

In order to reduce or eliminate possible interference with GPS receivers, UWB communications devices must comply with the limits shown in Table 2.9 and Figure 2.17.

Both the indoor and the handheld spectrum limits are identical and are displayed in Figure 2.17. This range covers the GPS operational band(s) between the frequencies 1,164–1,240 MHz and 1,559–1,610 MHz. The limit line is based on the RMS-detected signal measured with a resolution bandwidth of no less than 1 kHz, a video bandwidth greater than or equal to the resolution bandwidth, and a measurement (averaging time in the case of RMS detectors) time of 1 msec or less per frequency point.

As described in Section 2.4.6.5, all UWB emissions measurements performed at frequencies greater than 960 MHz are expressed in terms of their EIRP. GPS-band radiated-emissions data and the limit lines are expressed in units of dBm/kHz, provided the smallest unit of resolution bandwidth is used (unit of logarithmic power referenced to a milliwatt over a 1 kHz resolution bandwidth).

Frequency (MHz)	EIRP (dBm)	E (μV/m) @ 3m
1,164–1,240	–85.3	3.2
1,559–1,610	–85.3	3.2

Table 2.9 Spectral Line Emissions Limits Applicable to UWB Communications Systems

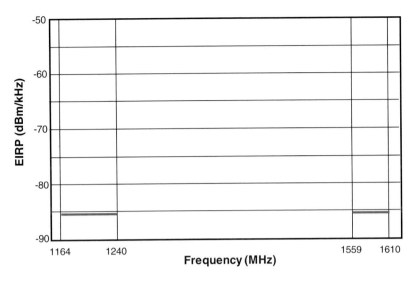

Figure 2.17 Spectral line emissions limits applicable to UWB communications systems.

2.4.7 Peak Power Emissions Limits

In addition to average power levels, the UWB peak power emissions above 960 MHz are also limited. The original FCC Part 15 rules were developed for relatively narrowband systems and, as such, are not appropriate or practical for measuring UWB technology. For example, initial Part 15 measurement procedures dictate the usage of a pulse desensitization correction factor. The application of this factor is not appropriate for very wideband systems. In fact, it may cause UWB systems to exceed the peak emission limits currently specified under the Part 15 rules [8]. With this in mind, the authorities created new measurement procedures and limits.

This limit is applicable to all UWB applications and is specified as an EIRP of 0 dBm within a 50 MHz measurement bandwidth centered on the frequency associated with the highest detected emission level above 960 MHz, f_M.

The resolution bandwidth must be equal to or between 1 MHz and 50 MHz (1 MHz ≤ RBW ≤ 50) when performing this measurement. Using a nominal value of 3 MHz for the resolution bandwidth is the recommended and preferred method. The reasoning behind this instrument setting will be explained later in this section.

The limit is specified using a 50 MHz RBW; therefore, if a smaller resolution bandwidth is used, the peak emissions EIRP limit (0 dBm/50 MHz) must be compensated to reflect that fact by applying a bandwidth scaling factor of $20\log_{10}(RBW/50 \text{ MHz})$. For example, the peak power limit can be expressed in a 3 MHz bandwidth as follows:

$$EIRP(3 \text{ MHz}) = EIRP(50 \text{ MHz}) + 20\log_{10}(3 \text{ MHz}/50 \text{ MHz})$$
$$= 0 \text{ dBm} + (-4.4 \text{ dB}) = -4.4 \text{ dBm}$$

During the open commentary portion of the rule-making procedure for UWB regulations, the application of a 50 MHz resolution bandwidth was a point of debate and discussion. Some of the questions revolved around the availability of a measurement receiver with a 50 MHz resolution bandwidth. Other questions concerned the accuracy or calibration methodology for such measurements. In the end, the FCC rationalized that most present-day spectrum analyzers or receivers have at least a 3 MHz resolution bandwidth with a matching 3 MHz video bandwidth. Using these bandwidths, most independent test laboratories would derive comparable measurements that the FCC could validate [8].

In the rules, it was decided that if the resolution bandwidth used is greater than 3 MHz, the application filed with the FCC must contain a detailed description of the test procedure, the calibration of the test setup, and the instrumentation used in the testing. This would be subject to the acceptance by the FCC [1].

2.4.7.1 Conducted ac Power Limits

Conducted ac line measurements are required for intentional radiators designed to be connected to the public-utility power grid. The measurement is to limit conducted noise pollution that can be coupled to the ac power lines and could cause interference in other aspects of the radiated and conducted environment. The measurements are made using a 50 µH/50Ω line impedance stabilization network. Compliance with the limits in Table 2.10 is based on the measurement of the RF voltage between each power line and ground at the power terminal.

Measurements to demonstrate compliance with conducted ac limits are not required for devices that are solely battery operated (e.g., handheld UWB devices); however, if the device has a battery charger, and the device can operate while it is charging, then the device must still demonstrate compliance with the conducted ac limits. (See Figure 2.18.)

2.4.8 Additional Technical Requirements Applicable to Indoor and Outdoor UWB Systems

CFR 47, Subpart F, § 15.521 contains additional items that both indoor and outdoor UWB systems must comply with. Some of the items have already been covered in other sections of this chapter since, thematically, they belong there. The remaining items are listed below with a brief explanation.

1. Restricted usage: UWB devices may not be employed for the operation of toys. Operation on aircraft, ships, or satellites is also prohibited.
2. Antenna requirements: An intentional radiator shall be designed to ensure that no antenna other than that furnished by the responsible party shall be used with the device. The use of a permanently attached antenna or an antenna that uses a unique coupling to the intentional radiator shall be sufficient to comply with

UWB Spectrum and Regulations   47

Frequency (MHz)	Emissions Limit (dBμV)	
	Quasi-peak	Average
0.15–0.5	66 to 56	56 to 46
0.5–5	56	46
5–30	60	50

Table 2.10 Conducted ac Power Line Emissions Limits Applicable to UWB Communications Systems

these regulations. Only the antenna for which an intentional radiator is authorized may be used with the intentional radiator.

3. System authorization. When a transmission system is authorized as a system, it must always be used in the configuration in which it was authorized.
4. Pulse gating. If pulse gating is used where the transmitter is quiescent for intervals that are long compared to the nominal pulse repetition interval, measurements must be made with the pulse train gated on. *Note:* This rule has been relaxed for UWB devices operating under certain constraints. Refer to Section 2.5 on the impact of the waiver [13].
5. Release from prohibition against damped wave emissions. UWB emissions are exempt from existing prohibitions applicable to damped wave emissions.
6. Appropriate calculations for exposure limits. These are generally applicable to all transmitters regulated by the FCC. At this time, UWB transmitters are excluded

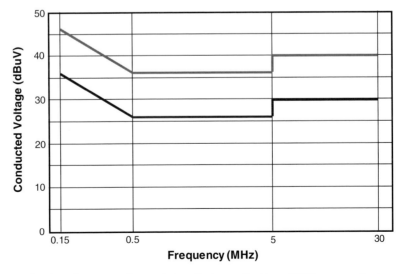

Figure 2.18 Conducted ac power line emissions limits applicable to UWB communications systems.

from submitting exposure studies or preparing environmental assessments other than for informative purposes; however, it is helpful to be prepared if these requirements are needed for future submittals. Familiarity with the exposure limits in 47 CFR, Part 1, Subpart 1, § 1.1310, and Part 2, Subpart J, § 2.1093, will be necessary.

2.5 UWB Waiver Impact on Technical Requirements

In August 2004, key representatives from the UWB industry filed a petition for waiver requesting that devices using frequency hopping, stepping, or sequencing be measured in their intended modes of operation. They argued that operating their radios in modes other than this did not illustrate a realistic usage scenario and would unfairly put their radios at a disadvantage by penalizing their performance.

Until the petition for waiver, the average emissions levels of any device were to be measured with the transmitter operating continuously and with any frequency hopping, stepping, or sequencing disabled. In addition, any gating was also to be disabled. The FCC requested that the radio be exercised in this overly conservative scenario based on two premises: the lack of established and proven measurement procedures for these radio types and the lack of information on the radio's interference potential.

In March 2005, the FCC released a grant of waiver [13]. The key points are as follows:

1. The requirement that measurements be made with the transmitter operating continuously with any frequency hopping, stepping, or sequencing disabled was removed. The device is to operate in its normal operating mode. Measurements of the average and peak emissions levels shall be repeated over multiple sweeps to ensure that the maximum signal is captured.

2. The waiver did not apply to systems that employ swept frequency modulation. Swept frequency systems were not addressed in the waiver and are outside the scope of discussion.

3. The gating requirements that measurements must be made with the pulse train gated on were waived. The emissions from all such systems shall be measured in their normal operating mode.

4. Operation of the waiver shall apply only to indoor and handheld UWB devices under 47 CFR, Subpart F, § 15.517 and § 15.519, that operate in the 3.1–5.03 GHz and/or 5.65–10.6 GHz frequency bands.

5. Because of the interference potential within the Microwave Landing System (MLS) and Terminal Doppler Weather Radar (TDWR) spectrum, fundamental emissions of the UWB device shall not be located within the 5.03–5.65 GHz band.

6. The system must comply with emissions levels under all possible operating conditions (for example, if the hopping sequence is changed).

7. The waiver shall not apply to the determination of the UWB bandwidth of the classification of a device as a UWB transmitter. The requirements of Subpart F, § 15.503(a, d) still apply based on measurements performed with any frequency hopping, stepping, or band sequencing stopped.

8. These rules are effective until the FCC finalizes a rule-making proceeding codifying these changes. Some of these restriction could be relaxed or constrained in the future if further information requires such as response. The delegation authority to modify the permit operation under this waiver has been placed in the control of the Office of Engineering and Technology.

2.5.1 Applying the Waiver

There could be different reasons to use the waiver when applying for regulatory acceptance. In order to be tested in their normal operating mode, all UWB indoor and handheld systems employing frequency hopping, stepping, or band sequencing will have to use the waiver to get through the regulatory process. In addition, since the gating restrictions are relaxed, transmission bursts can be optimized to provide the best propagation performance for the system. So, traditional baseband pulsed or pulse-modulated RF systems that were originally anticipated by the rules can apply for regulatory acceptance without the use of the waiver; however, they can also apply for regulatory compliance under the waiver if they want to take advantage of the loosened restrictions on gating (however, the radios will still have to go through a screening process to ensure that their interference potential has not been unduly increased by any optimized gating process). UWB submittals for communication systems are still in their infancy. This means that regardless of whether the waiver is used, all submissions will be carefully observed and put through a relatively diligent screening process involving the FCC laboratories. Since the FCC and the National Telecommunications and Information Administration (NTIA) are cooperating in this effort, expect delays in the first few sets of commercial UWB submissions. There must be a courteous coordination between the two organizations since the bandwidth allocated for UWB systems spans both the commercial and federal spectra.

2.6 International Regulatory Status of UWB Devices

Europe has traditionally been the leader regarding international harmonization and standardization of electromagnetic compatibility regulations [14] and measurement techniques. In the past 20 years more countries have harmonized to European norms or adopted European norms or trends as their foundation for EMC regulations than from any other source.

UWB regulations have deviated from this tradition. Initially, the foundation for UWB regulatory rules and measurements were proposed and adopted by the United States. These same rules have been used as foundations for other countries to begin their studies of implementing UWB technology. An example is Singapore's Infocomm Development Authority (IDA), which issues experimental licenses for the development of UWB based technologies.

Asia, in particular, has been on the fast track to study, implement and standardize rulings for new and emerging technologies. In the middle of September 2005 Japan's Ministry of Internal Affairs and Communications (MIC) completed the internal approval process of a proposed UWB spectrum mask. MIC is the government organization responsible for creating, among other things, Japan's fundamental information and communications structure. The latest version of the Japanese UWB spectrum mask, released in February 2006, is viewed as a successful first step towards global adaptation of a UWB spectrum policy.

In Japan the FCC's regulatory framework was adopted but the frequencies, power spectral densities and measurement methodologies underwent significant changes. Apart from slight frequency allocation differences, additional changes were made to the regulations that impact UWB systems. Several of the more important changes are noted below:

1. Coexistence and Mitigation: To protect the services of incumbent spectrum owners and to protect future services that may be introduced in band the concept of a "Detection-and-Avoidance" (DAA) technology was introduced. Japan has defined in band frequency zones, that are subsets of the overall UWB allocated frequency range. These sub bands are differentiated by their mitigation requirements. Some of these bands require DAA to be implemented. Some of these bands do not require DAA to be implemented. There is also a special case for one band that does not require DAA to be implemented until 4G services come on line in 2008.

2. Power Spectral Density Emissions Mask: There are several versions of reduced PSD emissions masks that have been published based on the applicability of DAA. All seem to imply that the original FCC mask limits do not provide enough protection for existing services and therefore should have lower emissions limits or more restrictive limits for the out of band emissions.

3. Measurement Methodologies: Due to extremely low power spectral density levels published in Japan's regulatory draft, conducted measurements will have to be used in lieu of radiated measurements over some frequency ranges. This is a large departure from the FCC's premise that the antenna be treated as an integral part of the device during measurements and cannot be separated from the radio. The effect the antenna has on the radio's spectrum is a detail that must be handled carefully if these regulations are to avoid abuse.

Japan's regulatory agency has agreed to review the process within three years to investigate issues that are still being put forth by developers. A few of the more popular requests have been; a slight frequency extension of the lower band edge to match the US standards; raising the average emission level in the higher band to compensate for system losses; and reducing the mitigation level requirements.

Other Asian and European Union proposed policies are on the horizon as well. During the time of this chapter's review period, there were still no a fully adopted standards or regulations which mandated the marketing and licensing of UWB products outside of the USA or Japan. However, it is clear that global industry support for UWB is gaining momentum.

Global UWB standards outside the United States will take a more conservative approach as described above. Presently, there is an effort to try and harmonize the regulations in Asia and Europe to reduce the amount of location specific adaptations of the technology. Since this process is extremely dynamic and changing quite rapidly, it is quite risky to predict the end result of international regulatory affairs concerning UWB within this chapter. It is probably a safe bet to assume that UWB will eventually make it into other markets. The question will be at what power spectral densities and with what operational caveats.

References

1. Title 47, Code of Federal Regulations (Washington DC: U.S. Government Printing Office, 2004).
2. Comité International Spécial des Perturbations Radioélectriques (CISPR), "CISPR 16-1-1: Specification for Radio Disturbance and Immunity Measuring Apparatus and Methods—Part 1-1: Radio Disturbance and Immunity Measuring Apparatus—Measuring Apparatus" (Geneva: IEC, 2003).
3. Institute of Electrical and Electronics Engineers (IEEE), *IEEE 100: The Authoritative Dictionary of IEEE Standards Terms* (New York: IEEE Press, 2000).
4. American National Standards Institute (ANSI), *ANSI C63.7: Guide for Construction of Open-Area Test Sites for Performing Radiated Emission Measurements* (New York: IEEE Press, 1992).
5. IEEE, *IEEE Std 149-1979: Standard Test Procedures for Antennas* (New York: IEEE Press, 1990).
6. Jones, Steve K. "A Methodology for Measuring Ultra-Wideband (UWB) Emissions to Assess Regulatory Compliance." Document No. USTG 1/8-31, March 9, 2004.
7. ANSI, *ANSI C63.4: Methods of Measurement of Radio-Noise Emissions from Low-Voltage Electrical and Electronic Equipment in the Range of 9 kHz to 40 GHz* (New York: IEEE Press, 2000).
8. First Report & Order in ET Docket No. 98-153, 17 FCC Rcd 7435 (2002), April 22, 2002.
9. "Spectrum Analyzer Basics." Agilent Application Note 150, 2005.
10. "Ultra-Wideband Communication RF Measurements." Agilent Application Note 1488, 2004.

11. McLean, James, et al. "Pattern Descriptors for UWB Antennas." *IEEE Trans. on Antennas and Propagation*, 53, no. 1 (January 2005): 553–559.

12. McLean, James, et al. "Interpreting Antenna Performance Parameters for EMC Applications—Part 1: Radiation Efficiency and Input Impedance Match." *Conformity* (July 2002).

13. Grant of Waiver in ET Docket No. 04-352: In Reply to Petition for Waiver of the Part 15 UWB Regulations Filed by the Multiband OFDM Alliance Special Interest Group, March 10, 2005.

14. European Telecommunication Standards Institute (ETSI), "Draft ETSI EN 302-065-1: Electromagnetic Compatibility and Radio Spectrum Matters (ERM); Short Range Devices (SRD) Using Ultra Wide Band Technology (UWB) for Communication Purposes in the Frequency Range 3.1 GHz to 10.6 GHz; Part 1: Technical Characteristics and Test Methods," Ref: DEN/ERM-TG31A-0012-1 (France: ETSI Press, 2003).

3

Interference and Coexistence

by Roberto Aiello

By its very nature, UWB must coexist with other systems in its broad bandwidth of operation. For instance, in the United States, the UWB frequency range for communication applications is 3.1 to 10.6 GHz (Figure 3.1), so it is operating in the same frequencies as popular consumer products, such as cordless telephones, digital TV, wireless LANs (WLANs), WiMAX, and satellite receivers, as well as commercial and governmental systems, such as navigational and meteorological radar. In other countries, it may also overlap with some emerging systems, such as third-generation (3G) or forth-generation (4G) wireless services.

The wireless industry, governments, and regulatory bodies have amassed a depth of research into the level to which UWB will interfere with other systems operating in the same frequency bands. Although the interpretations vary slightly and the type of UWB access technology plays some role in the level of interference, most agree that any UWB transmission must be limited to avoid the risk of harmful interference. Because there is no global standard or policy for allocation of bandwidth, each country that allocates spectrum will have a different list of potential "victim" receivers. For instance, in Europe, WiMAX services will use the 3 GHz band. As a result, any interference-mitigation techniques must be tuned for different markets.

Because it requires such an extensive range of spectrum, UWB has always been based on the principle of "underlay" [1], meaning that it must operate "under" other services in the same spectrum without causing harmful interference.

In order to ensure the effectiveness of the underlay principle, numerous governmental and organizational bodies, as well as manufacturers, have performed exten-

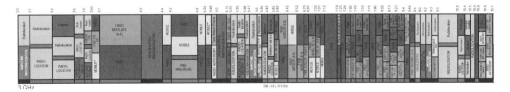

Figure 3.1 FCC frequency allocation chart for the UWB bandwidths.

sive testing and study to evaluate UWB's level of effect, if any, on specific incumbent services. Since it operates with a low-transmit power, the actual levels of interference are not as severe as some early estimates predicted. The remaining areas of concern have already been targeted by researchers in the UWB industry for mitigation techniques. This is a first step towards the concept of the "cognitive radio," defined as a radio that adapts its behavior based on external factors [2].

The technique showing the greatest promise to eliminate harmful interference in the presence of very sensitive services is Detection and Avoidance (DAA). In this approach, the UWB transmitter detects the presence of other active services (detection), then reduces its power in that specific band (avoidance). This "smart" technique allows UWB systems to operate across a continuous range of spectra, taking full advantage of the benefits of a wide frequency band. This technique shows great promise but has not been included in commercial products yet. This chapter describes some of the issues related to interference and coexistence.

3.1 Protecting Other Services in Band

To ensure its market and technical success, UWB cannot cause harmful interference with other services operating in the same frequency range. The challenge is that frequency ranges, applications, and philosophies differ between countries. For example, the United States recently allocated spectrum in the 3.6 GHz band for WiMAX, Europe allocated 3.4 to 4.2 GHz, and Japan has no WiMAX spectrum defined in the 3.5 GHz band. Achieving global spectrum harmonization would bring about greater competition, lower costs, and more potential for compliant devices to enter the market. However, this is not likely to happen, and the result is that the minimum requirements for protecting in-band services from interference are likely to differ by region for quite some time.

The first step to protecting other services, then, is to identify the areas of potential interference. Generally speaking, direct-sequence (DS) and impulse radio UWB devices produce a chip code sequence from subnanosecond pulses and combine it with the data sequence, resulting in a spread-spectrum signal with an approximate bandwidth of 1.5 GHz [3]. Multiband–orthogonal frequency division multiplexing (MB-OFDM) UWB devices transmit a 500 MHz signal that uses three bands below 5 GHz, switching bands every 312.5 ns. The interference potential of either UWB technology on broadband receivers depends on the nature of the victim receiver and the parameters of the transmitters. However, all UWB transmitters have the same effect on narrowband receivers (receivers whose bandwidth is lower than the impulse radio's pulse-repetition frequency or the OFDM switching rate).

Although the approaches differ, the end result is that UWB signals, regardless of the technology used, comprise a series of short bursts of RF. In contrast, services like WiMAX transmit data using time-based access technologies, with frame lengths varying between 2.5 and 20 ms and including a preamble, header, and user data. The challenge with predicting the impact of UWB devices on these time-based serv-

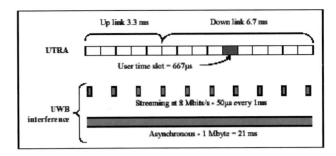

Figure 3.2 The probability of UWB signals to overlap with the signal for a WiMAX terminal. Source: A Technical Evaluation of the Effect of UWB on Broadband Fixed Wireless Access in the 3.4 GHz Band, *Indepen and Quotient, August 2005. p. 7.*

ices is that UWB systems behave differently depending on the application. For example, in a file-transfer application, the UWB device will send the data at the fastest available rate using consecutive frames. For streaming audio or video, it will transmit regular signal bursts at the highest available data rate. When these three types of transmissions are graphed together, there is obvious potential for interference (Figures 3.2 and 3.3) [4].

Typically, victim devices automatically boost power, switch to a more robust modulation scheme, or both in response to interference, the latter resulting in longer transmission time over the air. The net effect is increased "intercell" interference, warranting additional infrastructure build out. However, in a worst-case scenario, the UWB signal could still overpower the fixed broadband signal and block it. As a result, regardless of the amount of infrastructure, mitigation would be necessary in order to protect these other wireless services.

In another example, fixed satellite services (FSS) involve reception of very low signal levels from the satellites using high-gain, narrow-beamwidth antennas. Therefore, it is important to reduce the chances of interference from UWB devices that are likely to transmit signals directly into the main beam of these antennas.

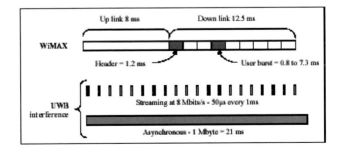

Figure 3.3 The probability of UWB signals to overlap with the signal for an ULTRA TDD terminal. Source: A Technical Evaluation of the Effect of UWB on Broadband Fixed Wireless Access in the 3.4 GHz Band, *Indepen and Quotient, August 2005. p. 7.*

Around the world, emission limits for UWB devices are being defined based on analyses such as these, and the services that require interference protection are being identified. The emission limits are currently being derived from numerous studies undertaken to classify the services that are most vulnerable in the bands under consideration.

In order to protect incumbent services adequately, it is important that regulators recognize the difference between specifying protection criteria for in-band and out-of band emissions [5]. From the UWB perspective, FSS operating in the 3.7–4.2 GHz band falls in the "in-band" services category, whereas personal communication services (PCS) or cellular services operating in the 1.7 GHz band are protected by the out-of-band emission limits set for UWB transmissions. This distinction becomes very important when we define levels for specific interference-mitigation techniques, such as DAA.

3.1.1 United States

When regulators allocate spectrum, they can follow various principles. For instance, in the United States, the FCC has a mandate to promote new technologies [6]. This guiding principle has the effect of creating more tolerance and promoting work with the proponents of new technologies, such as UWB, even though it has the potential to interfere with other services. UWB achieved initial approval in the United States for unlicensed communication devices operating in the 3.1–10.6 GHz space in February 2002 [7]. The FCC defined a UWB device as "one that emits signals with −10 dB bandwidth greater than 500 MHz, or greater than 20 percent of the center frequency." The FCC UWB regulations set upper limits on the amount of power that can be radiated across 3.1 to 10.6 GHz and out of band; this is known as a *mask*.

Specifically, the FCC rules include emission limits of −41.3 dBm/MHz maximum average EIRP spectral density with 1 ms integration time and 0 dBm peak power in a 50 MHz bandwidth. This is a very low power level; in fact, it is the same limit unintentional radiators must meet to get FCC certification. In the case of the FCC requirements concerning UWB, protection for out-of-band systems actually exceeds that required of other radio transmitters approved for use in the spectrum (Figure 3.4).

The FCC's recommendations were grounded in facts collected by numerous studies undertaken over the years to assess the compatibility of UWB. For instance, in 2000, the U.S. Department of Commerce/National Telecommunications and Information Administration (NTIA) released a study titled *Assessment of Compatibility between UWB Devices and Selected Federal Systems*, which aimed to assess the potential impact of UWB devices on federal equipment operating between 400 and 6,000 MHz, which includes 18 bands and a total of 2,502.7 MHz of restricted spectrum. As a result of its analysis, the NTIA calculated the maximum permissible average EIRP density in a 1 MHz bandwidth (average EIRP, dBm/MHz rms) for the systems under study, which included the Next Generation Weather Radar (NEXRAD)

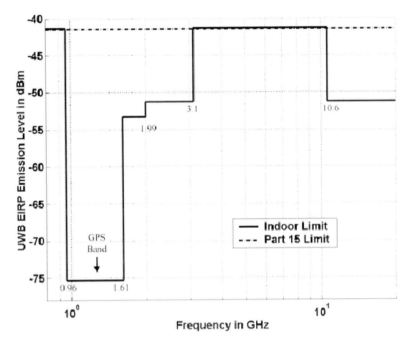

Figure 3.4 FCC Spectrum mask for UWB in the United States.

system, Air Route Surveillance Radar (ARSR)–4, Airport Surveillance Radar (ASR)–9, RF altimeters, the Air Traffic Control Radio Beacon System (ATCRBS), distance measuring equipment (DME), the Microwave Landing System (MLS), the Search and Rescue Satellite (SARSAT) Land User Terminal (LUT), a 4 GHz Earth station, a Terminal Doppler Weather Radar (TDWR) system, and a shipboard maritime radio navigation radar.

In addition, the NTIA determined the minimum separation distance between a receiver and a UWB device with an average EIRP spectral density of –41.3 dBm/MHz (RMS). The study considered the effects of a single UWB emitter on one receiver and the aggregate effects of several UWB emitters on one receiver. Throughout the study, the UWB devices were assumed to overlap completely the bands used by the equipment being assessed. The analytical results developed were then compared with measurements made at NTIA's Institute for Telecommunication Sciences (ITS) in Boulder, Colorado, and field measurements made at the Federal Aviation Administration facilities in Oklahoma City, Oklahoma.

This study concluded,

> This report shows that operation of UWB devices is feasible in portions of the spectrum between about 3.1 and 5.650 GHz at heights of about 2 meters with some operating constraints. Operations of UWB devices below 3.1 GHz will be quite challenging and any policy developed will need to consider the results of the analyses of interactions of GPS and UWB systems underway at NTIA and other facilities.

While the study showed that aggregate UWB interference can be a significant factor to receiving systems under ideal propagation conditions, a number of mitigating factors must also be taken into account that may reduce or eliminate these aggregate effects. There are also numerous mitigating factors that could relax restrictions on operation of UWB devices below 3.1 GHz.

A companion NTIA study titled *Assessment of Compatibility between UWB Systems and Global Positioning Systems Receivers* was undertaken to define the maximum allowable UWB EIRP levels that can be tolerated by GPS receivers in order to ensure their operations. These EIRP levels were then compared to the emission levels derived from the limits specified in 47 CFR, Part 15.209 to assess the applicability of the Part 15 limits to UWB devices.

The study concluded,

For those UWB signals examined with a PRF of 100 kHz, maximum possible EIRP levels between –73.2 and –26.5 dBW/MHz are necessary to ensure electro-mechanical compatibility (EMC) with the GPS applications defined by the operational scenarios considered within this study.

. . .

The data collected in this assessment demonstrates that when considered in potential interactions with GPS receivers used in applications represented by the operational scenarios considered in this study, some of the UWB signal permutations examined exceeded the measured GPS performance thresholds at EIRP levels well below the current Part 15 emission level. Likewise, other UWB signal permutations (e.g., the 100 kHz PRF UWB signals) only slightly exceeded, and in some cases did not exceed, the measured GPS performance thresholds when considered in potential interactions with GPS receivers defined by the operational scenarios considered as part of this study.

This is consistent with other studies performed by both UWB proponents and critics [8].

3.1.2 Europe

The International Telecommunications Union (ITU) [9] has a mandate to coordinate the operation of telecommunication networks and services. Its general approach tends to support strongly existing services and entities who have already paid for licenses and who expect a certain level of quality and performance.

In Europe, Ofcom, the independent regulator and competition authority for UK communications industries with responsibilities spanning television, radio, telecommunications, and wireless communications services, has been very active in recommending and promoting regulations for the acceptance of UWB. Its approach has been to focus on maximizing financial gain for all parties. It must be noted that this objective is very different from the FCC's and the ITU's and it is expected to lead to different recommendations.

Early in its analysis, Ofcom commissioned Mason Communications and DotEcon, independent consultants, to prepare a preliminary financial analysis of the effect of UWB devices in the United Kingdom. Specifically, the study compared the net benefits to consumers of using UWB-enabled devices (rather than alternative technologies) when considering the expected external costs in terms of spectral interference with other radio services.

The consultants analyzed four scenarios:

1. The FCC indoor mask (the UWB emission limits for UWB indoor communications applications adopted by the U.S. FCC)
2. Two versions of the draft ETSI UWB mask for UWB indoor communications systems currently being considered within the European Conference of Postal and Telecommunications Administrations (CEPT) and ETSI.
3. The current version, with a Power Spectral Density (PSD) of –65 dBm/MHz at 2.1 GHz ("ETSI mask" in Figure 3.5).
4. A revised version with a tighter PSD of –85 dBm/MHz at 2.1 GHz. This is termed either the ("proposed Ofcom revision to the ETSI mask" in Figure 3.7)
    - Lower band only—restricting UWB PAN transmissions between 3 and 5 GHz
    - Upper band only—restricting UWB PAN transmissions between 6 and 10 GHz

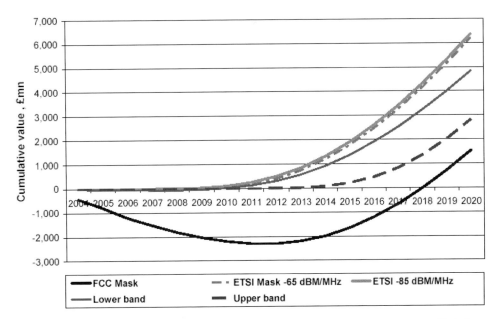

Figure 3.5 Ofcom financial analysis of the net benefits to consumers using UWB-enabled devices (rather than alternative technologies).

The main finding of the study was that "UWB has the potential to make a substantial contribution to the UK economy, generating about £4 billion (discounted) in value over the next 15 years. For the period to 2020, net private benefits exceed external costs under all the regulatory scenarios considered (however, in the case of the FCC mask, a positive net value is not achieved until 2020, and significant external costs are present in preceding period) [10].

With the establishment of likely financial gain for Europeans, Ofcom turned its attention to the issue of potential interference issues. During the period between 2002 and 2005, numerous studies were undertaken to determine the potential effect of UWB on other services in the 3.1–10.6 GHz band. In a consultation document released in January 2005, Ofcom reviewed studies performed on behalf of the FCC, the ITU, and CEPT [11]. Ofcom concluded,

> In Ofcom's assessment, if we allow UWB devices subject to a restrictive mask there is unlikely to be a significant increase in the risk of harmful interference in most cases. We have also noted that UWB has been allowed in the United States. If we do nothing, there are risks that UWB equipment will arrive in the United Kingdom in any case, and that the U.S. mask, which we believe not to be optimal for the United Kingdom, will become the de facto world standard. In light of this, Ofcom considers it to be particularly important to develop an appropriate European approach to UWB as soon as practicable. [12]

In 2005, Ofcom also commissioned two consultants, Indepen and Quotient, to undertake an independent analysis specifically of the potential effects of UWB on incumbent broadband fixed-wireless-access systems in the United Kingdom. The analysis team modeled likely scenarios using Monte Carlo simulation, statistically taking into account the variability of device locations and the bursty nature of their transmissions. The team made a series of assumptions to develop their simulation

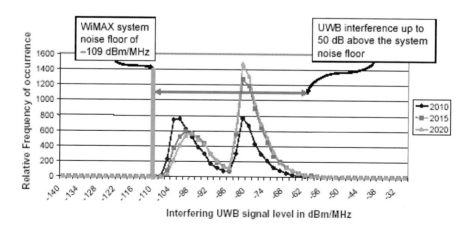

Figure 3.6 The relative frequency of occurrence is shown as a function of the interfering signal level experienced at the BFWA antenna. Results for the office environment are illustrated for each of the three years, 2010, 2015, 2010.

Interference and Coexistence 61

Frequency Range (GHz)	Spectral Mask (dBm/MHz)	DAA Requirement
3.1 to 4.2	−41.3	DAA required at −70 dBm/MHz
4.2 to 4.8	−41.3	No DAA required until 2010
6 to 10.6	−41.3	No DAA requirement

Table 3.1 Spectrum Allocations in Europe: Proposed Mask

and included a "bias towards conservatism" so that, if anything, it would overstate the risk of interference [13].

This study aimed to examine closely the effects of UWB devices and fixed-wireless broadband devices collocated in an office environment. The results showed that interference tended to center on two peaks, one at −78 dBm (which reflects the UWB devices close to the primary PC) and one at −96 dBm (which results from other UWB devices spread around the room). These results were graphed in relation to the noise floor for a WiMAX (Figure 3.6) system and demonstrated that interference from UWB had the potential to be more than 50 dB above the minimum operating level of the system. This means that additional protection is required to reduce its effect on those systems.

As a result of the Ofcom work and the UWB industry studies, European regulators have recommended the use of a spectral mask (Table 3.1) [14]. In this case, DAA is required in the 3.1–4.2 GHz band, while no DAA is required until 2010 in the 4.2–4.8 GHz band, and no DAA is required in the 6–10.6 GHz band. DAA means that the power in a specific frequency needs to be reduced when

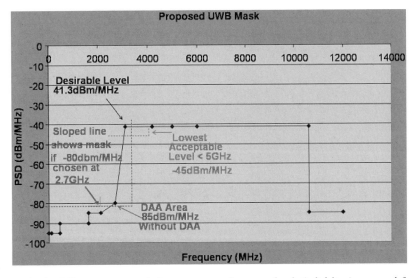

Figure 3.7 In the Ofcom recommended spectrum mask, power levels (solid line) proposed for Europe are equivalent to United States regulations, but the DAA requirements (dotted line) are more stringent than those in Japan.

62 Chapter 3

another service is detected. Interference mitigation can be achieved with a notch induced dynamically into the transmit spectrum or by lowering the transmit power.

Figure 3.7 shows that power levels (solid line) proposed for Europe are equivalent to U.S. regulations, but the DAA requirements (dotted line) are more stringent than those in Japan [15].

3.1.3 Japan

In August 2005, the Japanese Ministry proposed that UWB would span 3.4 to 10.25 GHz but could not interfere with transmission in the future 3.6 GHz WiMAX band (Figure 3.8). The ministry concluded that its country's services required more protection than U.S. services in the same frequency spectrum and specified that the transmitter must notch power down by −70 dBm/MHz in the lower bands.

Figure 3.9 shows a summary of the proposed mask for Japan. It specifies power levels equivalent to those of the United States in the 3.4–4.8 GHz band, but with an additional requirement for DAA with mitigation level set to −70 dBm/MHz to

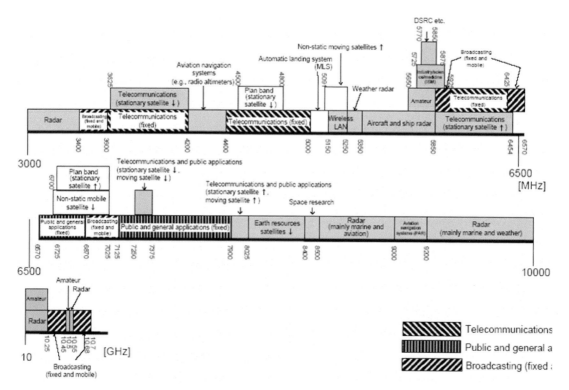

Figure 3.8 Japanese frequency allocation.

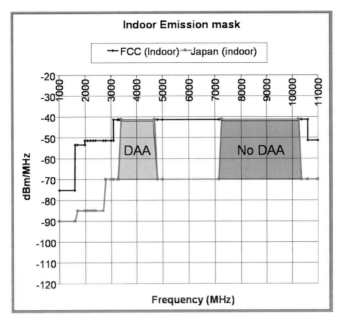

Figure 3.9 Proposed Japanese spectrum mask.

protect broadcast services and future cellular services, also known as fourth-generation (4G) services. In addition, transmissions in the 3.1–3.4 GHz band are not allowed in order to protect radio astronomy services (RAS) operating at 3.3 GHz.

3.1.4 International Big Picture

Some regions have proposed breaking the available spectrum into chunks, providing a noncontinuous band for UWB operation. This is certainly not ideal for UWB and would negatively impact its performance.

Japanese and European administrations have proposed a different set of protection criteria based on the services that operate in their respective regions. They have recently added the need for DAA, which is similar to the Dynamic Frequency Selection (DFS) scheme used by 5 GHz Wireless Access Systems (WAS) in Europe to protect primary services.

Table 3.2 sums up the current situation in the three areas leading the world with the most UWB interest and activity, the United States, Japan, and Europe.

In the end, there is no best approach that works for all countries, but governments, regulators, and industry need to find the right balance that best benefits the public. From a UWB industry perspective, it is beneficial to find a common framework so that the same technology can be used for different countries with only minor modifications.

Category	Region	Frequency Bands and Additional Requirements
Full band	United States	3.1 to 10.6 GHz –41.3 dBm/MHz
Low band	European Union	3.1 to 4.95 GHz –41.3 dBm/MHz with either protection mechanism: • Using DAA with transmit level reduced to –70 dBm/MHz in presence of other services that require protection • Restricting duty cycle to a maximum of 5 percent over 1 second and 0.5 percent over 1 hour 4.2 to 4.8 GHz No limitation until 2010
	Japan	3.4 to 4.8 GHz –41.3 dBm/MHz with DAA with transmit level reduced to –70 dBm/MHz in presence of other services that require protection
High band	European Union	6 to 8.5 GHz –41.3 dBm/MHz
	Japan	7.25 to 10.25 GHz –41.3 dBm/MHz

Table 3.2 Summary regulatory situation in US, Europe and Japan.

3.2 Ensuring Coexistence

Industry experts and standards bodies agree that, to be successful, UWB transmitters must detect other services that are active in the vicinity and avoid interfering with them. For instance, if a WiMAX network is nearby, the UWB transmitter should check the WiMAX spectrum to see if there is a high-powered signal. If it detects a signal, then it should avoid that piece of spectrum for the current transmission.

Taking this idea further, if there is a signal in the WiMAX base station band, for instance, but there are no receivers interested in connecting to it, then the UWB transmitter should not be required to notch out that piece of spectrum. This approach certainly adds to the complexity of the DAA scheme but prevents unnecessary performance degradation for the UWB system and optimizes the use of the spectrum.

3.2.1 Technical and Market Challenges

The first requirement of a UWB system is that it not interfere harmfully with other services, either now or in the future. The challenge is that UWB's multigigahertz span will surely include future spectrum allocations and usage models that we cannot predict today. And, it must be recognized that physical separations may be insufficient to prevent interference with some of these future services.

Several approaches have been studied in order to mitigate interference. Some proposed approaches include controlling transmit power, switching off UWB devices that are close to primary PCs in office environments, limiting the duty cycle of UWB devices, and moving UWB devices to higher frequencies. These methods have been found to be insufficient in mitigating the interference at the required level [16]. A 2005 Ofcom study concludes that "by attenuating UWB emissions by 40 dB, the interference impact can be reduced to around 4 percent of network site costs for the high BFWA market forecast and around 1 percent of network site costs for the moderate BFWA market forecast. Attenuation by 50 dB appears to remove all effects of UWB interference."

Regulations for UWB are likely to continue to vary from place to place around the world, undoubtedly necessitating spectrum mask compromises. In addition to spectrum masks, regulations in Japan, Europe, and other countries will also likely require DAA.

3.3 Detection and Avoidance

DAA is the leading technique for solving UWB-related interference issues. In this approach, the UWB transmitter detects whether another active service is present (detection) and the likelihood that the interference is high, then it reduces power in that specific band (avoidance). This "smart" technique allows UWB systems to operate across a continuous range of spectrum, taking full advantage of the benefits of a wide frequency band. A UWB and WiMax scenario, shown in Figure 3.10, provides a good example of DAA usage. At present, WiMax is the only in-band (3.4–3.8 GHz) operational service in Europe. In the worst case, a WiMax client node may be on the edge of a base station cell and may experience serious fading of the downlink signal transmitted by the base station. This may seem to be an extreme case; however, regulators need to find a good balance between conservative cases and realistic scenarios.

It is possible that a low power UWB radio operating in close proximity to a WiMax receiver raises the noise floor by an amount sufficient to cause performance

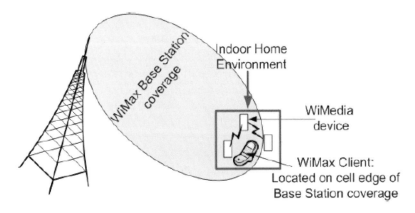

Figure 3.10 Potential areas of interference for a UWB system.

degradation to the receiver. Modern wireless technologies are adaptive in nature and a loss in performance observed at the WiMax receiver will probably cause the WiMax transmitter to fall back to a more robust modulation scheme with lower overall throughput, or to step up the transmit power to maintain the overall system performance. This scenario would not be effective if the WiMax receiver cannot detect the base station at all. Cell boundary interference scenarios could also be reduced by reducing cell size, but only with considerable additional infrastructure costs.

The remainder of this chapter describes how this method can be applied to a MB-OFDM system (see Chapter 8).

For instance, DAA is expected to be used in Japan in a two-step process. In Step 1, Detection and Coarse Avoidance, the system will run the detection algorithm and effectively shut off the band containing the victim services and continue using the bands that are free of victim services. Currently, there is a low probability that victim services will be in two bands simultaneously, so the UWB system's performance should not be affected.

Step 2 in this proposed process is known as Detection and Fine Avoidance. In this case, specific avoidance in narrow bands is achieved by notching out the minimum amount of spectrum is used.

The impact of these regulations on impulse radios is likely to be very costly at present; therefore, it appears that the frequency of operation for those radios will be shifted to higher frequencies where DAA is not required.

3.3.1 Detection

One of the challenges of detection is to pick up the very low victim service signal power in the presence of outgoing UWB transmissions and other noise sources. The scheme must also be flexible enough to accommodate both present and future services in the bands of interest, which means that the technique should be independent of the service.

Figure 3.11 shows a simplified block diagram of the detection process that can be implemented in UWB radios based on the fast Fourier transform (FFT) [17]. Real conditions generally complicate the problem, and it is necessary to include out-of-band emissions from other services, RF interference, and even UWB transmissions from neighboring nodes in the equation. All of these must be incorporated into the algorithm to achieve accurate detection of services.

3.3.2 Avoidance

After a potentially interfering signal is detected, a number of solutions are available to address it. Current options include transmit power control (TPC), frequency notching, and active interference canceling.

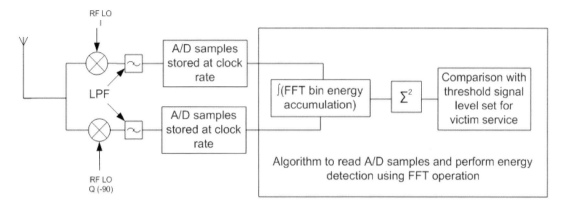

FFT bin size equivalent to Integration time
Energy includes UWB noise and WiMax signal components

Figure 3.11 A simplified block diagram of the detection process that can be implemented in UWB radios based on the Fast Fourier Transform (FFT) functionality available in the PHY architecture.

TPC. TPC is based on the principle of transmitting the minimum amount of power required to transmit the information. This is a useful tool, employed with basic techniques, to increase the level of coexistence. It is widely used in many wireless systems.

Frequency notching, or "tone nulling." This technique involves inserting zeroes at the FFT stage to achieve notches in the transmit spectrum. Theoretically, a five-bit digital-to-analog converter (DAC) would achieve a 30 dB notch; however, practical implementation and system nonlinearities are likely to restrict the notch to 15 to 20 dB.

Figure 3.12 shows an example of tone nulling for narrowband service operating at 4 GHz [18]. In this case, we have the following characteristics:

- The 128 tone OFDM transmitter "zeros out" six adjacent tones near the detected service.
- Out of 128 tones, 122 are still available.
- Notch depth is approximately 15 dB.
- It can be done with all-digital operation (no analog filtering required).

Advanced techniques. More advanced techniques are being developed that achieve deeper notches using digital techniques. The example shown in Figure 3.13 [19] is a simulation with a 30 dB notch across a 200 MHz band achievable using 50 in-band tones (50 × 4.125 = 206.25 MHz) and only six additional tones. In this example, the UWB average power level is –41.25 dBm/MHz, and the targeted mitigation level is –70.0 dBm/MHz; so the required mitigation depth of 29 dB can achieve deeper notches in the transmit spectrum with a modest increase in power

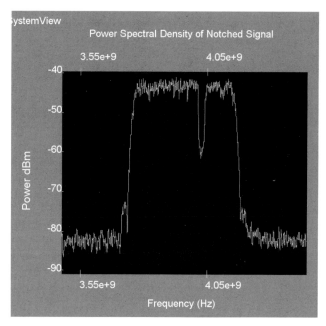

Figure 3.12 An example of tone nulling for narrowband service operating at 4 GHz.

consumption, hardware complexity, and cost due to more computation at the transmitter [19].

Figure 3.14 shows another example, where a 28 dB notch depth is achieved using an actual UWB transmitter.

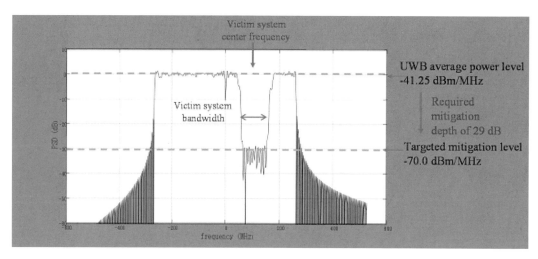

Figure 3.13 In addition to notching, a fixed number (maximum of six required for a notch width of 200 MHz) of neighboring tones are computed and combined with the in-band tones to achieve a deeper notch.

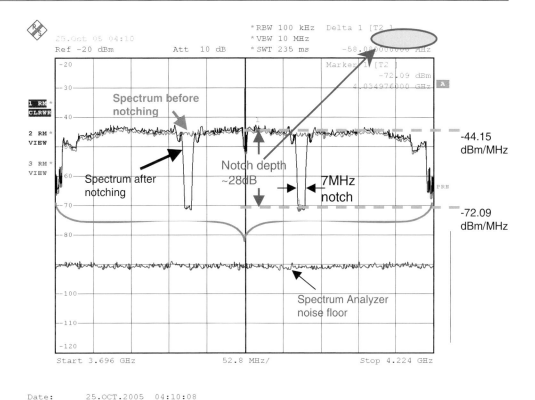

Figure 3.14 An example of tone nulling measured on a real system in the laboratory.

3.4 Responding to Changing Needs

In order to stay current with regional requirements for interference mitigation, it is important that the UWB technology be as flexible as possible. For instance, OFDM transmitters generate the transmit signal in the frequency domain and then transform it in the time domain. So, if it is necessary to avoid a specific frequency, an OFDM transmitter can send a lower power signal in that specific frequency.

An impulse, or DS-UWB, transmitter requires an additional frequency-domain mechanism, such as a notch filter, to avoid specific frequencies. The disadvantage of a notch filter is that it adds cost to the unit and is not easily tunable to other frequencies. It is likely that impulse radios will initially operate at higher frequencies, where DAA is not required, in Japan and Europe until digital-notching techniques that allow the required protection have been developed.

OFDM waveforms are generated in the frequency domain and offer some inherent advantages for meeting the varying regulations in different regions. For instance, a key feature of a long-lived UWB PHY is that its radiated spectrum shape can be controlled through software in near real-time in order to respond to local regulations and time-varying local conditions. By ensuring that this shaping is only performed as much as necessary (and no more than necessary), the negative effects on UWB performance can be minimized.

It is reasonable to think that new technologies will be developed in the future with even more spectrum flexibility to allow even better adaptation to global coexistence requirements. In the meantime, UWB is very likely to be one of the first technologies deployed that allows dynamic adaptation to interference.

3.5 Finding the Balance

It is true that there are challenges with regard to finding the right place for UWB in the range of wireless services and devices available in international markets. Fortunately, industry leaders have worked out reliable DAA techniques to ensure the quality of incumbent services while making room for UWB services that will enhance communications services and options for the public. Systems based on OFDM are likely to focus initially in the lower bands, which include DAA techniques, while impulse radio–based systems are likely to focus initially on the upper bands where DAA is not required.

The proposals from the European Commission and the Japanese Ministry are important definitive steps towards UWB deployment worldwide. As regulatory bodies and governments around the globe continue to define the requirements for UWB, manufacturers can work to find the common platforms, techniques, and deployment scenarios that will best support these global services in a cost-effective manner.

References

1. Jim Lovette was the first to propose the concept in 2000.
2. FCC 05-57, ET Docket No. 03-108.
3. Robert A. Scholtz and Moe Z. Win. "Impulse Radio." (Paper given at IEEE PIMRC '97, Helsinki, Finland).
4. Indepen and Quotient. "A Technical Evaluation of the Effect of UWB on Broadband Fixed Wireless Access in the 3.4 GHz Band." *(August 2005)*: 5.
5. Out-of-band emission limits are fixed and, hence, can be realized using external filters with sharp roll off; whereas in-band emission limits, especially the dynamic changes in spectral characteristics, cannot be implemented using external filters.
6. Communications Act of 1934, 47 U.S.C. 157, § 7, New Technologies and Services.
7. FCC. "First Report and Order, Revision of Part 15 of the Commission's Rules Regarding Ultra Wideband Systems." ET Docket 98-153 (2002).

8. Aiello, R., G. Rogerson, and P. Enge. "Preliminary Assessment of Interference between Ultra-Wideband Transmitters and the Global Positioning System: A Cooperative Study." *Proceedings of the 2000 Institute of Navigation National Technical Meeting*, Anaheim, California (January 2000): 28–35.
9. The ITU, headquartered in Geneva, Switzerland, is an international organization within the United Nations system where governments and the private sector coordinate global telecom networks and services.
10. Ofcom. "Ultra Wideband Consultation Report." (January 2005): 20.
11. CEPT was established in 1959 by 19 countries, which expanded to 26 during its first 10 years. Original members were the incumbent monopoly-holding postal and telecommunications administrations. CEPT's activities included cooperation on commercial, operational, regulatory, and technical standardization issues. In 1988, CEPT decided to create ETSI, into which all its telecommunication standardization activities were transferred.
12. Ofcom. "Ultra Wideband Consultation Report." (January 2005): 34.
13. Indepen and Quotient. "A Technical Evaluation of the Effect of UWB on Broadband Fixed Wireless Access in the 3.4 GHz Band." August 2005.
14. Electronic Communications Committee (ECC). "ECC Decision of 2006 on the Harmonized Conditions for the Use of UWB Devices below 10.6 GHz." September 2005.
15. UKTG1/8(05)26. "Input Document for Discussion at the UK TG3 Preparation." August 2005.
16. Indepen and Quotient. "A Technical Evaluation of the Effect of UWB on Broadband Fixed Wireless Access in the 3.4 GHz Band." August 2005.
17. Multiband OFDM Physical Layer Specification Release 1.0.
18. IEEE P802.15-05/425r0, September 2004.
19. ECC/TG3 TG#10_05r0. "UWB Interference Detection Technology." June 2005.

4

UWB Antennas
by James S. McLean and Heinrich Foltz

Broadband antennas have been around for many decades and are used extensively, especially for reception. In some regards, traditional broadband antennas satisfy the requirements for commercial UWB systems; thus, many of the existing design approaches can be borrowed intact. Nevertheless, despite its designation as UWB, the proposed commercial UWB radio concept with its frequency range of 3.1 to 10.6 GHz differs significantly from traditional wideband, short-pulse applications, such as radar.[1] Moreover, UWB antennas for portable consumer electronics and mobile communications applications have different requirements than those intended for fixed-point communications or broadband testing; as a consequence, many of the conventional frequency independent and log-periodic designs are not suitable. For example, size constraints often dominate all other engineering considerations, precluding the use of many traditional UWB designs. Also, for devices for which the placement and orientation cannot be specified a priori, a nominally omnidirectional, low-gain pattern is generally desirable since the use of a high-gain antenna would make placement and orientation of the units critical.

The traditional parameters used to describe the performance of antennas may not be the most convenient description for UWB applications. While, in principle, complete knowledge of the antenna is contained in the impedance response and the gain, polarization, and phase patterns as functions of frequency, UWB communications system designers may prefer to consider ringing time, energy efficiency, energy gain, pulse fidelity, or the transfer function for a complete link including propagation model.

This chapter is divided into four parts. The first reviews basic restrictions on the bandwidth and size of antennas, which limit the range of performance that can be obtained even in principle. The second part introduces antenna transfer functions and shows the signal modification inherent in a wireless link. The third part covers UWB antenna performance descriptors and how they can be related back to conventional antenna parameters. Finally, the fourth part gives a brief survey of recent UWB antenna designs, emphasizing small, low-gain antennas that may be suitable

---

[1] For an excellent and thorough treatment of the topic of short-pulse antennas, the reader should refer to [1].

for mobile and portable electronic devices. This part also discusses general considerations in the choice of balanced versus unbalanced antennas and the importance of the structure to which the antenna is connected.

Although much of the material presented applies to broadband operation in general, we emphasize issues and antenna designs relevant to the 3.1–10.6 GHz range allocated by the FCC for commercial UWB devices in the United States. We assume the reader to be familiar with basic antenna terminology and principles. Several excellent textbooks [2–4] are available for those needing an introduction to antenna fundamentals.

4.1 Fundamental Limitations on Performance

Maxwell's equations and circuit theory lead to fundamental restrictions on the performance of antennas that are restricted in size [5–12]. Size, gain, impedance bandwidth, and efficiency are interrelated such that an improvement in one parameter necessitates a sacrifice in one or more of the others. These trade-offs become particularly apparent when the antenna is electrically small. Strictly speaking, electrically small antennas are defined as those that fit into a sphere one electrical radian ($\lambda/2\pi$) in radius. This definition is rather arbitrary, and, in practice, trade-offs of size versus other parameters may become a serious design issue for antennas that are larger than the strict definition.

4.1.1 Fundamental Physical Limits on the Impedance Bandwidth of Antennas

An antenna that is strictly electrically small over its entire operating range cannot simultaneously be highly efficient and provide the impedance bandwidth required for a typical UWB application. A much more common scenario is an antenna that is electrically small at the lower end of its operating frequency range and of moderate electrical size at the upper end. The bandwidth predictions derived for electrically small, narrowband antennas are not appropriate without modification.

Narrowband antennas are typically modeled as second-order bandpass networks. The equivalent networks models for spherical wave functions have the form of highpass filter. Similarly, the equivalent networks for many simple canonical antennas, such as the tapered dipole, are highpass in nature. Thus, for an antenna that operates over a frequency range extending from frequencies at which it is electrically small to those in which it is of moderate size, it is more appropriate to treat the matching problem as that of a highpass filter.

4.1.1.1 Radiation Q

The radiation quality factor Q of an antenna is 2π times the ratio of the maximum instantaneous nonpropagating stored energy in either the electric or magnetic field (whichever is greater) to the power radiated in one cycle. It is defined for sinusoidal inputs and, thus, is a Q function of frequency. It can be expressed as

$$Q = \frac{2\omega W_s}{P_{rad}} \quad (4.1)$$

where W_s is the time-averaged, nonpropagating stored energy in either the electric or magnetic field, whichever is greater, and P_{rad} is the time-averaged radiated power. The actual Q of realistic antennas is very difficult to compute directly from this definition since it is not straightforward to separate nonpropagating fields from propagating fields in either measurements or numerical simulations. More frequently, it is deduced from equivalent circuit models.

However, although it is difficult to compute for any particular structure, it can be shown [5–12] that the Q is subject to fundamental, unavoidable constraints regardless of how the antenna is designed. For an antenna that fits into a sphere of radius a and radiates an electromagnetic field composed of exclusively TM or exclusively TE spherical modes, the constraint is that

$$Q \geq \frac{1}{ka} + \frac{1}{(ka)^3} \quad (4.2)$$

Such a spherical-mode representation is typical of an antenna, such as a linear or tapered dipole, that exhibits a radiation pattern that is isotropic in one plane. If the antenna is allowed to have an arbitrary polarization pattern (e.g., a combination of TE and TM modes), the constraint is

$$Q \geq \frac{1}{ka} + \frac{1}{2(ka)^3} \quad (4.3)$$

For $ka \ll 1$, the lower limit on the radiation Q for an antenna with an arbitrary polarization pattern is approximately one-half that of an antenna occupying the same volume and radiating exclusively TM or TE modes.

If the antenna is constructed over an infinite ground plane (i.e., a monopole or one of its derivatives), a sphere encompassing both the antenna and its image is used. If the antenna operates in the vicinity of some other arbitrary object, such as a finite ground plane or, perhaps, the casing of a radio, the characteristics of the image can be very complicated. In accurately assessing the effective size of the antenna, it is absolutely essential that the entire image be accounted for.

4.1.1.2 Approximate Bandwidth Limits: Wheeler Formula

The simple relationship of Q to bandwidth

$$B = \frac{f_0}{Q} \qquad (4.4)$$

is only valid when the system in question is narrowband second order, and the bandwidth is defined by the half-power point. It is certainly not true in any sense in the case of a realistic antenna with 130 percent impedance bandwidth. More accurate relationships between the maximum feasible bandwidth and the Q have been given by Harold Wheeler [9, 13], and by R. M. Fano [14].

For the simple case of a lossless, electrically small antenna with an arbitrary lossless matching network, Wheeler has provided the following relationship:

$$\left[\left(\frac{f_0}{f_L}\right)^3 - \left(\frac{f_0}{f_H}\right)^3\right] \ln|1/\Gamma_{max}| < \frac{3\pi}{Q} \qquad (4.5)$$

where Q is specified at f_0, f_L is the lower band edge, f_H is the upper band edge, and Γ_{max} is the maximum acceptable reflection coefficient in the passband. This inequality is a fundamental limit that can only be reached, in principle, with an infinite order matching network. For a "double-tuned matching network," the limitation on bandwidth is only $2/\pi$ as wide.

Despite its limitations, the Wheeler relationship shows a number of considerations for UWB antennas:

- First, it is possible for f_H to be infinite and still satisfy the inequality, indicating that a highpass response can, at least in principle, be obtained from a finite-sized antenna.
- Second, in an antenna with an octave or more bandwidth, the f_H term is relatively small, so size limitations are essentially determined by the lower band edge only.
- Third, the obtainable bandwidth is closely related to the reflection coefficient specification. System designers should realize that specifying a very low Voltage Standing Wave Ratio (VSWR) will impact the feasible antenna size, and vice versa.

For antennas that are very small at the lower band edge (e.g., $ka < 0.2$), the Wheeler bandwidth formula can be combined with the approximation $Q \approx 1/(ka)^3$ to obtain an effective limit on the feasible reflection coefficient. At some point in the band, the magnitude of the reflection coefficient will rise to at least

$$\Gamma_{max} > \exp\left\{\frac{-24\pi^4 a^3}{c_0^3} \frac{f_L^3 f_H^3}{f_H^3 - f_L^3}\right\} \quad (4.6)$$

or, in the highpass case, to

$$\Gamma_{max} > \exp\left\{\frac{-24\pi^4 a^3 f_L^3}{c_0^3}\right\} \quad (4.7)$$

These limits apply to antennas based on TM or TE modes alone. For antennas that can use an arbitrary combination of TM and TE, $Q \approx 1/2(ka)^3$, and the equations become

$$\Gamma_{max} > \exp\left\{\frac{-48\pi^4 a^3}{c_0^3} \frac{f_L^3 f_H^3}{f_H^3 - f_L^3}\right\} \quad (4.8)$$

in the bandpass case, and

$$\Gamma_{max} > \exp\left\{\frac{-48\pi^4 a^3 f_L^3}{c_0^3}\right\} \quad (4.9)$$

in the highpass case.

Note that these limits are highly optimistic, in addition to being only approximately applicable, as mentioned above. Trying to reduce the reflection coefficient to less than Γ_{max} in one portion of the band will necessarily force it to be higher than Γ_{max} in another portion of the band; with a finite order matching network, the match will be significantly poorer. For a 10 dB return loss and the 3.1–10.6 GHz UWB band, (4.6) for the TE- or TM-only case gives a minimum feasible size of a = 7.6 mm. For 20 dB return loss, a = 9.6 mm. Allowing arbitrary polarization, as in (4.8), reduces the minimum feasible sizes to a = 6.0 mm and a = 7.6 mm for the 10 dB and 20 dB return-loss cases, respectively. Again, these limits are approximate and overly optimistic. More accurate limits based on Fano's formulation are discussed in Section 4.1.1.3.

Of course, the radiation Q limit can be evaded by having ohmic loss in the antenna, either through unavoidable loss in the materials used or losses deliberately added through resistive loading. However, under normal conditions, the degradation in radiation efficiency will more than cancel out the improvement in matching efficiency. Resistive loading may still be appropriate if the driving circuit is highly sensitive to mismatch, and it may help to control distortion of the pulse shape; however, it will usually not be effective in improving the total amount of energy transferred from the circuit to space.

4.1.1.3 Bandwidth Limitations Based on Fano Theory

As noted above, the entire concept of radiation Q is closely tied to the idea of a narrowband, second-order circuit and, thus, only applies to UWB antennas in a general sense. An alternative approach that does not rely on the radiation Q is that of Fano [14]. Fano addressed the problem of broadband impedance matching to the input of networks consisting of an arbitrary lossless network with a resistive termination (the equivalent circuit models for spherical modes have exactly this form). His results are presented in the form of integral relationships that the reflection coefficient must satisfy.

The derivation of Fano's integral relationships is involved. For the case of the circuit models for spherical modes in free space, they are

$$\int_0^\infty \frac{1}{\omega^{2(k+1)}} \ln\left(\frac{1}{|\Gamma(\omega)|}\right) d\omega = (-1)^k \frac{\pi}{2} F_{2k+1}^0 \qquad (4.10)$$

for $k = 0, 1, 2, \ldots n$, where n is the order of the spherical Hankel function and where

$$F_{2k+1}^0 = A_{2k+1}^0 - \frac{2}{2k+1} \sum \left(\frac{1}{\lambda_{ri}}\right)^{2k+1} \qquad (4.11)$$

The A_{2k+1}^0 are determined by the spherical mode in question, and once a bounding sphere has been selected, they are out of our control. The λ_{ri} are zeroes of the reflection coefficient of the antenna/matching network combination, which lie in the right half-plane. Other than this restriction, they can, in principle, be selected arbitrarily through the design of the antenna structure and any additional matching circuitry that is added. In practice, connecting the structure to the values is difficult; it is more important that the summation in the first equation work out to be real and positive (positive because the zeroes are, by definition, in the right half-plane and real because they occur in conjugate pairs).

For the lowest order mode TM_{01}, we have $n = 1$ and $k = 0$ and 1, and the coefficients are $A_1^0 = 2a/c_o$ and $A_3^0 = -4a^3/3c_o^3$. This leads to the simultaneous constraints:

$$\int_0^\infty \frac{1}{\omega^2} \ln\left(\frac{1}{|\Gamma(\omega)|}\right) d\omega = \frac{\pi}{2}\left[\frac{2a}{c_o} - 2\sum \frac{1}{\lambda_{ri}}\right] \qquad (4.12)$$

and

$$\int_0^\infty \frac{1}{\omega^4} \ln\left(\frac{1}{|\Gamma(\omega)|}\right) d\omega = \frac{\pi}{2}\left[\frac{4a^3}{3c_o^3} + \frac{2}{3}\sum\left(\frac{1}{\lambda_{ri}}\right)^3\right] \qquad (4.13)$$

The integrals can be thought of as a "matching area." Again, the λ_{ri} are set by the design of the antenna and the matching network, and the summations are real and positive. Suppose that one wanted to optimize use of the "matching area" represented by the integrals. Γ should have a magnitude of 1 outside the desired band and be constant, $|\Gamma| = \Gamma_0$, within the band, so that

$$\left(\frac{1}{\omega_L} - \frac{1}{\omega_H}\right) \ln\frac{1}{\Gamma_0} = \frac{\pi}{2}\left[\frac{2a}{c_o} - 2\sum\frac{1}{\lambda_{ri}}\right] \qquad (4.14)$$

$$\frac{1}{3}\left(\frac{1}{\omega_L^3} - \frac{1}{\omega_H^3}\right) \ln\frac{1}{\Gamma_0} = \frac{\pi}{2}\left[\frac{4a^3}{3c_o^3} + \frac{2}{3}\sum\left(\frac{1}{\lambda_{ri}}\right)^3\right] \qquad (4.15)$$

The values of λ_{ri} are numerically optimized to find the minimum feasible size for a given reflection coefficient, or vice versa. For a 10 dB return loss and a 3.1–10.6 GHz band, the minimum feasible size is $a = 8.5$ mm; for 20 dB, it is $a = 10.9$ mm. Figure 4.1 shows the relationship between size and return loss. Note that these sizes are larger than those predicted by the Q-based formulas in the previous section and are presumably more accurate because no small-antenna approximation is used in the derivation. For small, broadband monopole or dipole antennas with greater than one octave bandwidth and VSWR < 2, the current state of the art is about 1.5 times as large as the theoretical fundamental limit.

Derivations for higher-order spherical modes are similar in principle but more difficult to solve analytically. However, the limitations on the higher-order modes are more severe, and utilizing antennas that radiate them can only increase the minimum feasible size or degrade the best attainable reflection coefficient. If it is acceptable for the antenna to radiate a combination of TE and TM modes, it is clear that the size and bandwidth constraints are somewhat relaxed, as was mentioned in the previous section. We are not aware of a solution of this case based on the Fano formulation; however, a reasonable conjecture would be that the feasible sizes would be reduced by a factor of approximately $\sqrt[3]{2}$.

80 Chapter 4

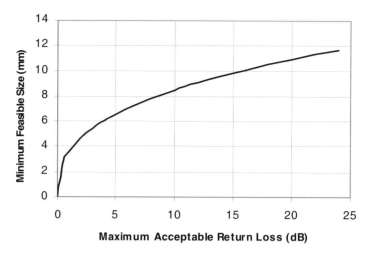

Figure 4.1 Size versus return loss constraints for 3.1–10.6 GHz antenna based on Fano formulation. Based on antennas radiating TM_{01} or TE_{01} modes alone; antennas radiating both TE and TM waves can be somewhat smaller; antennas radiating higher order modes must be larger.

4.1.2 Limitations on Gain versus Size

In many UWB antennas intended for mobile or portable applications, the pattern is not a critical parameter, and the gain is not critical as long as the efficiency is high. However, in those applications where a controlled directional pattern is desired, fundamental size constraints again become important. There is no theoretical limit to the maximum gain obtainable for a given size at a single frequency, but increasing the gain makes the impedance-bandwidth-versus-size constraints described in the previous section much more severe.

4.1.2.1 Maximum Antenna Gain, Supergain

If one expands the radiation fields of an antenna in spherical harmonics, it can be shown [15] that the maximum possible gain is given by

$$G_{max} = n^2 + 2n \quad (4.16)$$

where n is the highest-order spherical Hankel function used in the expansion. A rough idea of the required size can be obtained by noting that spherical mode Q increases rapidly when $ka < n$, thus making broadband gain difficult to obtain. Setting $ka = n$ gives the *normal gain*, defined [15] as

$$G_{norm} = (ka)^2 + 2ka \quad (4.17)$$

Antennas with a higher gain than G_{norm} for a given size are referred to as *supergain* antennas. Equation (4.17) can easily be solved to give the required minimum size a_{norm} to avoid the supergain condition for a desired gain G:

$$a_{norm} = \frac{\sqrt{1+G}-1}{k} \quad (4.18)$$

Equation (4.17), and hence Equation (4.18), does not take into account the desired reflection coefficient tolerance and uses a subjective definition of high Q. Also, it is well known that small, low-gain antennas can (and often must) operate under supergain conditions but that supergain becomes much more difficult to obtain for high-gain antennas. It is reasonable to assume that low-gain antennas will be supergain in the lower part of the UWB passband.

4.1.2.2 Relation of Gain to Q and Bandwidth

Some further (but still approximate) insight can be obtained by relating the maximum gain G_{max} to the bandwidth restrictions discussed earlier. Application of the Fano theory of Section 4.1.1.3 to the case of a combination of higher-order modes is difficult, and the Wheeler expression of Section 4.1.1.2 does not apply directly to higher-order circuits. However, exact expressions for the Q of higher-order modes are available in [11]. The expressions are somewhat lengthy and will not be repeated here, but calculations of the required electrical size ka for different values of Q are given in Table 4.1.

We can approximately relate the Q to the achievable bandwidth by making four assumptions: (1) as before, only TE or TM modes will be used, (2) the Q will be dominated by the behavior of the highest-order mode, (3) the antenna cannot be tuned for a given level of reflection unless the Q is less than or equal to that required for the TM_{01} mode according to the exact Fano theory, and (4) the Q should be taken at the lower band edge since this is where tuning is most critical.

n	G_{max}	Required ka for $Q = 26.0$	Required ka for $Q = 7.75$	Required ka for $Q = 4.24$
1	3	0.351	0.552	0.707
2	8	1.01	1.40	1.70
3	15	1.79	2.35	2.83
4	24	2.63	3.38	4.08
5	35	3.52	4.46	5.46

Table 4.1 *Required Electrical Size to Obtain a Given Q Value for Different Mode Numbers. (In application to a UWB antenna, ka should be taken at the lower band edge.)*

From the data in Section 4.1.1.3 for the TM_{01} mode, a return loss of 3 dB requires $ka > 0.351$, a loss of 10 dB requires $ka > 0.552$, and a loss of 20 dB requires $ka > 0.707$, all taken at the lower band edge, and assuming a 3.4:1 bandwidth (e.g., 3.1 to 10.6 GHz). The values for ka are not strongly affected by the upper-to-lower frequency ratio as long it is substantially more than one octave. Using (4.2), the Q for the TM_{01} mode for these values of ka is 26.0, 7.75, and 4.24, respectively. We then find, in Table 4.1, the ka values for the higher-order modes to get these same Q values. In using the same Q values for higher values of n, we are making the assumption (stated above as item 3) that tuning these modes will be at least as difficult as tuning the TM_{01} mode for the same Q.

The results of the table verify the well-known rule of thumb that $ka \approx n$ and the normal gain G_{norm} are practical limits for large, high-gain antennas, but that substantially higher gain is feasible and eventually required as the size is decreased.

It should be stressed that the values given in Table 4.1 for ka are exact for the stated Q, but the relation between Q and obtaining the specified return loss over a wide bandwidth is an estimate (except for $n = 1$). Thus, the table should be not taken as a fundamental limit on size (e.g., the greatest lower bound); however, it is unlikely that smaller antennas could be obtained. The bandwidth used to compute the sizes in Table 4.1 was 3.4:1, and the resulting values of ka should be taken at the lower band edge.

4.2 Transfer Characteristics: The Antenna as a Filter

It is often convenient to view antenna-to-antenna transmission as a linear, reciprocal, passive two-port network interconnecting two radios. Such a two-port model is useful when considering the effects of noise and computing channel capacity or bit error rate. In the static case, no matter how complicated the propagation scenario is, a single, passive two-port network can represent the antenna-to-antenna transmission; that is, the port-to-port transmission can be considered as a linear time-invariant (LTI) network. This is true even when mutual coupling between antennas is not negligible [16]. Indeed, the most rigorous derivation of the frequency-domain transfer function between two fixed antennas is obtained using spherical-mode expansions and does not require the assumption of reciprocity (although the system is certainly reciprocal) [16]. In a dynamic situation, in which either an antenna or a scattering object in the vicinity moves or is reoriented, the parameters of the equivalent two-port network vary with time. Thus, the antenna-to-antenna transmission can be modeled as a lossy, linear, two-port network that, due to its frequency response, distorts signals.

Obtaining a "flat" response for the equivalent two-point network is not simply a matter of having antennas with a flat gain versus frequency. As a simple example, consider three equivalent forms of the basic Friis transmission equation [2, p. 79]:

$$\frac{P_R}{P_T} = G_T G_R \left(\frac{\lambda}{4\pi R}\right)^2 = G_T G_R \left(\frac{c_0}{2\omega R}\right)^2 = G_T A \frac{1}{4\pi R^2} \quad (4.19)$$

where A is the effective aperture of the receive antenna. Constant gain with frequency is often assumed to be a desirable characteristic of a UWB antenna; however, it is apparent from the second form that two constant-gain antennas in a free-space environment give a response that drops as the square of frequency. To get a flat response, some other combination is needed; for example, the third form shows that the combination of a constant-gain antenna at one side and a constant-aperture antenna at the other side will yield frequency-independent power transfer.

4.2.1 Limitations

Before continuing, we should note that the transfer function approach necessarily includes a propagation model. Most analytical derivations of transfer functions, including the one presented here, assume free-space conditions with spherical spreading (i.e., conditions under which the Friis transmission equation applies). It would be fair to say that in mobile and indoor communications work, there are few situations in which the Friis equation is even roughly valid. The orders of both the frequency variation and the variation with distance are likely to be different, not to mention the occurrence of random variations due to multipath. Nevertheless, the simple transfer functions are a useful starting point in that they help to make clear what types of signal distortions are inherent in a wireless link and to distinguish antenna effects from propagation effects. The remainder of this section will explore these concepts in more detail.

4.2.2 Derivation of Antenna-to-Antenna Transfer Functions for UWB Applications

4.2.2.1 Effective Length

In deriving an antenna-to-antenna transfer function, it is natural to begin with the so-called effective length or height of an antenna [17, 18]:

$$V_{rxOC}(\theta,\phi,\omega) = -\vec{h}_e(\theta,\phi,\omega) \cdot \vec{E}^{inc}(\theta,\phi,\omega) \quad (4.20)$$

Here, $V_{OC}(\theta,\phi,\omega)$ is the open-circuit voltage produced across the antenna terminals by the incident electric field, $\vec{E}^{inc}(\theta,\phi,\omega)$ is the incident electric field, and $h_e(\theta,\phi,\omega)$ is the complex vector effective height or length of the antenna. The terms *effective height* and *effective length* are used interchangeably in the literature. Effective height was originally coined for monopoles. In any case, the parameter has units of length and calls to mind the integrating nature of a linear receive antenna. It is possible to derive a relationship between the radiated electric field of an antenna and the base current using the Lorentz reciprocity theorem:

$$\vec{E}^{rad}(\theta,\phi,\omega) = -\underbrace{j\omega}_{\text{time differentiation}} \underbrace{\frac{\eta_0}{c_0}}_{} \underbrace{I_a}_{\text{base current}} \underbrace{\vec{h}_e(\theta,\phi,\omega)}_{\text{effective length}} \underbrace{\frac{e^{-j\beta R}}{4\pi R}}_{\text{spherical spreading}} \quad (4.21)$$

where I_a is the base current or current at the input terminals of the antenna. In some cases, this relationship is taken as defining effective length, and the first equation is then derived using reciprocity. In any case, if one of the two equations is taken as the definition of the effective length, the other follows from reciprocity. It is useful to split up the expression into the factors shown here in order to call attention specifically to several different mechanisms. Notice that if the effective length is constant, a time waveform is differentiated upon transmission if one considers the input signal to be the input current. However, it is important to remember that the effective length as defined by the previous equation is seldom constant with frequency for any antenna with the exception of a short dipole. In addition, the relationship between the base current and the source voltage is complicated by impedance mismatch.

4.2.2.2 Radiated Field

The base current of the transmitting antenna, therefore its radiated field, is related to the open-circuit voltage of the source via the combined generator and antenna impedances:

$$\vec{E}^{rad}(\theta,\phi,\omega) = -jV_g(\omega)\underbrace{\frac{\eta_0}{Z_g+Z_a(\omega)}}_{\text{dimensionless}} \underbrace{\beta\vec{h}_e(\theta,\phi,\omega)}_{\text{antenna field factor}} \frac{e^{-j\beta R}}{4\pi R} \quad (4.22)$$

Here, the expression for the antenna field factor defined in [19] is bracketed. Note that it is dimensionless, as is the preceding factor. The two taken together are defined later in Section 4.3.3.3 as a flatness function for the antenna.

4.2.2.3 Transfer Functions

There are several variations of definitions for transmit and receive transfer functions [19–26]. The antenna field factor as defined above was given by Lamensdorf and Susman [19] for transmit antenna transfer function. The complex antenna factor

(CAF) given in [22] can be thought of as a receive transfer function. A. Shlivinski, E. Heyman, and R. Kastner [23] give a detailed discussion of decomposition of the transfer function into components due to the receive and transmit antennas. However, the transfer function for antenna-to-antenna transmission can be derived in terms of vector effective length without defining any other quantities. When the antennas are constructed from reciprocal materials and the intervening medium is also reciprocal, the forward and reverse transfer scattering parameters will be identical functions of frequency. Thus, a set of n fixed antennas in a static environment (i.e., one with no moving scatterers) can be represented as a lossy n-port scattering network.

From the expression for the received voltage, (4.20), and for the transmitted electric field, (4.21), we can immediately write

$$\frac{V_{recOC}}{I_a} = Z_{21} = Z_{12} = j\omega \frac{\eta_0}{c_0} \vec{h}_e^{tx}(\theta_{tx}, \phi_{tx}, \omega) \vec{h}_e^{rx}(\theta_{rx}, \phi_{rx}, \omega) \frac{e^{-j\beta R}}{4\pi R} \quad (4.23)$$

This is the complex transfer impedance of the two-port network. It is a function of the positions and orientations of both antennas. Note that four angles are required to specify the geometrical relationship between the two antennas. When mutual coupling is neglected in the computation of the reaction on each antenna, the three unique entries in the impedance matrix representation of the two antennas are given by the two antenna input impedances and this transfer impedance; that is, the diagonal entries are given by the respective input impedances. Of course, other two-port representations are valid. In particular, the admittance matrix can often be directly computed in numerical simulations employing impressed electric field sources.

One particular pair of definitions for transmit and receive antenna transfer functions has gained general acceptance [24–26]. Here we assume the generator and load (receiver) impedances are real and equal to a system impedance Z_0.

For this special case, the transmitted electric field in terms of the normalized incident voltage, $a(\omega)$, is

$$\vec{E}^{rad}(\theta, \phi, \omega) = -ja(\omega) \frac{\eta_0 \sqrt{Z_0}}{Z_0 + Z_a^{tx}(\omega)} \beta \vec{h}_e(\theta, \phi, \omega) \frac{e^{-j\beta R}}{2\pi R} \quad (4.24)$$

since $a(\omega) = \frac{V_g(\omega)}{2\sqrt{Z_0}}$.

The transmit antenna transfer function can be defined as

$$\vec{H}_{tx}(\theta,\phi,\omega) = \frac{\dfrac{\vec{E}^{rad}(\theta,\phi,\omega)}{\sqrt{\eta_0}}}{\dfrac{V_+(\omega)}{\sqrt{Z_0}}} 2\pi R \; e^{j\beta R} = \frac{\sqrt{\eta_0 Z_0}}{Z_0 + Z_a^{tx}(\omega)} \vec{h}_e(\theta,\phi,\omega) \quad (4.25)$$

where $V_+(\omega) = \dfrac{V_g(\omega)}{2}$ is the incident voltage at the antenna input port.

The transmit transfer function has units of length and relates the normalized incident voltage to the normalized transmitted electric field. The transmit antenna transfer function is a vector function of frequency, as well as the angular position of the observation point.

The normalized transmitted voltage at the receiver port is

$$b(\omega) = -\left(\frac{\sqrt{Z_0}}{Z_0 + Z_a^{rx}(\omega)}\right) \vec{h}_e(\theta,\phi,\omega) \cdot \vec{E}^{inc}(\theta,\phi,\omega) \quad (4.26)$$

Now the receive transfer function can be defined as H_{rx}, where

$$b(\omega) = -\vec{H}_{rx}(\theta,\phi,\omega) \cdot \frac{\vec{E}^{inc}(\theta,\phi,\omega)}{\sqrt{\eta_0}} \quad (4.27)$$

Thus,

$$\vec{H}_{rx}(\theta;\phi;\omega) = \vec{h}_e(\theta,\phi,\omega)\left(\frac{\sqrt{Z_0 \eta_0}}{Z_0 + Z_a^{rx}(\omega)}\right) \quad (4.28)$$

The receive transfer function also has units of length and is a vector function of frequency, as well as the angular position of the observation point.

Finally, an antenna-to-antenna transfer scattering parameter,

$$S_{21} = \frac{b(\omega)}{a(\omega)},$$

can be derived as follows:

$$S_{21} = j\left(\frac{\eta_0}{Z_0 + Z_a^{tx}(\omega)}\right)\beta \vec{h}_e^{tx}(\theta_{tx},\phi_{tx},\omega) \cdot \vec{h}_e^{rx}(\theta_{rx},\phi_{rx},\omega)\left(\frac{Z_0}{Z_0 + Z_a^{rx}(\omega)}\right)\frac{e^{-j\beta R}}{2\pi R} \quad (4.29)$$

$$S_{21}(\omega) = j\omega \vec{H}_{tx}(\theta_{tx},\phi_{tx},\omega) \cdot \vec{H}_{rx}(\theta_{rx},\phi_{rx},\omega)\frac{e^{-j\beta R}}{2\pi R} \quad (4.30)$$

where the transfer scattering parameter is a function of the relative position and orientation of both antennas.

In order to examine in detail several of the factors in the transfer function, one can take the inverse Fourier transform of the received voltage:

$$b(\omega) = S_{21}(\omega)a(\omega) = j\omega \vec{H}_{tx}(\theta_{tx},\phi_{tx},\omega) \cdot \vec{H}_{rx}(\theta_{rx},\phi_{rx},\omega)\frac{e^{-j\beta R}}{2\pi Rc_0}a(\omega) \quad (4.31)$$

$$b(t) = \frac{1}{2\pi}\int_{-\infty}^{+\infty} j\omega \vec{H}_{tx}(\theta_{tx},\phi_{tx},\omega) \cdot \vec{H}_{rx}(\theta_{rx},\phi_{rx},\omega)\frac{e^{-j\beta R}}{2\pi Rc_0}a(\omega)e^{j\omega t}d\omega \quad (4.32)$$

First, a time differentiation is apparent:

$$b(t) = \frac{1}{2\pi}\frac{\partial}{\partial t}\left[\int_{-\infty}^{+\infty} \vec{H}_{tx}(\theta_{tx},\phi_{tx},\omega) \cdot \vec{H}_{rx}(\theta_{rx},\phi_{rx},\omega)\frac{e^{-j\beta R}}{2\pi Rc_0}a(\omega)e^{j\omega t}d\omega\right] \quad (4.33)$$

Second, a time delay is implicitly contained in the complex exponential:

$$b(t + R/c_0) = \frac{1}{2\pi}\frac{1}{2\pi Rc_0}\frac{\partial}{\partial t}\left[\int_{-\infty}^{+\infty} \vec{H}_{tx}(\theta_{tx},\phi_{tx},\omega) \cdot \vec{H}_{rx}(\theta_{rx},\phi_{rx},\omega)a(\omega)e^{j\omega t}d\omega\right] \quad (4.34)$$

The transfer function between various canonical antennas can be determined analytically. For example, the transfer function between two electrically small linear dipole antennas has been given in [21, 27, 28].

In the case of pulse transmission, it is possible to define energy "link loss" [21] as

$$L_{\text{link}} = \frac{W_{\text{rec}}}{W_{\text{in}}} \quad (4.35)$$

where W_{in} is the energy accepted by the transmit antenna, and W_{rec} is the antenna delivered to the receiver; that is, the link loss is really a measure of energy transfer.

Here,

$$W_{\text{in}} = \frac{1}{2\pi} \int_0^\infty \frac{|V_G(\omega)|^2 \operatorname{Re}(Z_T(\omega))}{|Z_T(\omega) + Z_G(\omega)|^2} d\omega \quad (4.36)$$

and

$$W_{\text{rec}} = \frac{1}{2\pi} \int_0^\infty \frac{|V_L(\omega)|^2}{\operatorname{Re}(Z_L)^2} d\omega \quad (4.37)$$

where the received voltage is computed from the transfer function. The energy accepted by the transmit antenna is also the available source energy multiplied by the matching efficiency given earlier. Of course, like the matching efficiency, the link loss is a function of the pulse shape. In [21], the link loss between two electrically short dipoles is analytically derived for Gaussian and monocycle or Gaussian doublet pulse shapes. For both cases, the link loss is computed as a function of receiver input resistance and is found to exhibit a maximum when the receiver resistance is on the order of a few thousand ohms. We note that the value of this maximum, while similar for both pulse shapes is not truly independent of pulse shape.

4.2.2.4 Antenna-to-Antenna Transfer Function Example

Two useful canonical transfer function examples are the transmission between two conical monopoles and two triangular monopoles mounted on a large ground plane. The large ground plane essentially removes any effects of imbalance to simulate perfectly balanced dipoles without the effects of feed transmission lines. In Figure 4.2, the conical monopole is shown mounted in the ground plane. The included angle of the monopole is 60°, and the height of the monopole is 3 cm. The monopole is fabricated from aluminum with a copper insert to allow direct soldering to an SMA connector. In this manner, no other support mechanism is necessary, and the antenna comes very close to realizing a true conical shape. Figure 4.3 shows the entire two-antenna configuration. It is useful to make the measurements in an

Figure 4.2 Conical monopole mounted on large ground plane.

Figure 4.3 Monopoles mounted on ground plane situated in anechoic chamber.

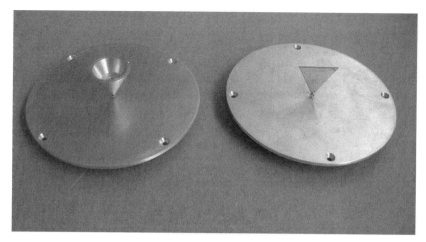

Figure 4.4 Conical and bowtie monopoles. The height of each antenna is 3 cm.

anechoic chamber in order to reduce the effects of reflections found in a typical room. Figure 4.4 shows the individual conical and triangular antennas with their respective inserts. The height of the triangular monopole is also 3 cm to facilitate comparison. The triangular monopole is fabricated from two-sided FR-4 circuit board with both sides soldered to the SMA connector. The FR-4 circuit board was

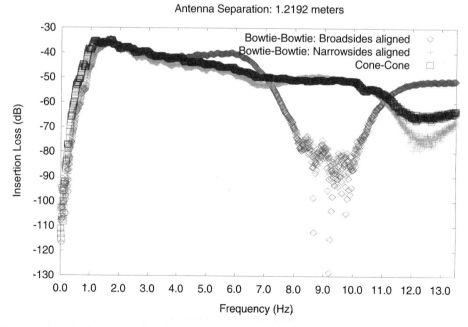

Figure 4.5 Insertion loss and return loss for a pair of conical monopoles. The ground plane is 1.2192m × 2.4384m and the spacing between monopoles is 1.2192m. The cone angle is 60°.

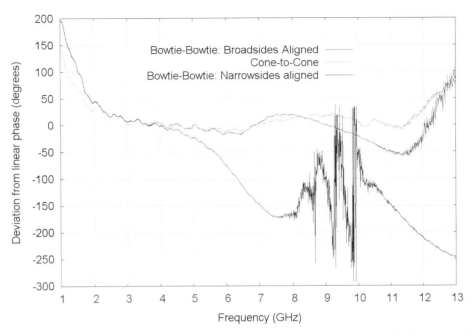

Figure 4.6 Deviation from linear phase of the port-to-port transfer functions for the monopoles described in Figure 4.5.

thought to be stiffer and flatter than sheet metal stock. Like the conical antenna, no support mechanism other than the solder joint was used. Figure 4.5 shows the measured return loss and insertion loss for the pair of conical monopoles and the pair of triangular monopoles. The roll-off at the low end is due to impedance mismatch. The variation of insertion loss with frequency is less than the 12 dB/octave variation one would expect from a constant-gain antenna because the gain of the monopole increases slightly with frequency up to the point at which the monopole becomes about two-thirds of a wavelength high. The small quasiperiodic ripple in the insertion loss is due to reflections from the edge of the ground plane. It is interesting to note that the pattern of the triangular monopole is complicated above 6 GHz, as is evidenced by the markedly different transfer functions obtained by rotating the triangular monopoles 90°.

Figure 4.6 shows the deviation from the linear phase of the port-to-port transfer functions for the monopoles described in Figure 4.5.

4.3 UWB Antenna Performance Parameters

The desirable parameters for a UWB antenna depend strongly on the application, propagation environment, and modulation scheme. In particular, one must consider the following:

- whether the modulation and equalization scheme require a *specific* received pulse shape, a *consistent* received pulse shape, or *any* shape within a given time limit, or the system is not pulse-based at all;
- whether the link is fixed, portable, or mobile;
- whether the link has a dominant line-of-sight path or is dominated by multipath.

The modulation scheme determines whether antenna-induced dispersion is an issue. For example, in an orthogonal frequency division multiplexing (OFDM) system, the linearity of the phase response over the entire UWB band, that is, its freedom from dispersion, is not important as long as the bandwidth occupied by each of the subcarriers is sufficiently small. Furthermore, OFDM is normally intended for operation in multipath environments, in which the multipath delay spread would normally dwarf any time delays created by the antennas. Therefore, in such a system, one would be most concerned with the *energy efficiency* and *energy pattern* of the antenna.

On the other hand, impulse radio systems require that the separateness of the pulses be maintained. In a simple system based on detecting the presence or absence of pulses, the shape of the pulse may be of little importance, but it is important to minimize the time spreading of the pulses. In this case, a measure of the time distribution of energy, such as T_{99} or the *time-windowed energy efficiency*, would be of interest.

In a system where information is transmitted through a symbol set consisting of different pulse waveforms, the pulse shape is important. The pulse shape is also important in high-resolution imaging and locating applications. If a specific shape is required, the *fidelity* parameter or similar measure is needed. If arbitrary but fixed equalization is available, any shape may be acceptable, but the shape should be consistent with direction. In this case, the *correlation pattern* or *correlated energy gain pattern* are appropriate.

The distinction between fixed, portable, and mobile links is also critical. (Here we define *fixed* as a situation where antennas can be intentionally aligned, such as a roof-mounted point-to-point link, and *portable* as a situation where there is no motion, but the antenna alignment will be random, such as a consumer-deployed home wireless router.) The most obvious impact on the antenna is that portable and mobile systems generally require omnidirectional antennas; however, another distinction between portable and mobile units is that the adaptive equalization problem is considerably more difficult in the mobile case. If the adaptive algorithm in a pulse radio system cannot keep up in real time with the motion, it effectively becomes a fixed scheme and the *correlated energy gain* becomes an important parameter.

4.3.1 Efficiency

The matching efficiency, energy radiation efficiency, and energy efficiency are measures of how well the antenna transports energy from transmitter to a radiated field, or from a radiated field to a receiver. They are applicable to virtually all UWB applications.

4.3.1.1 Definition of Matching Efficiency

The matching efficiency e_m is a measure of what fraction of the available pulse energy is delivered from the source to the antenna or vice versa, as opposed to being reflected. That is,

$$e_m = \frac{W_d}{W_a} \quad (4.38)$$

where, in the case of a transmit antenna, W_d is the energy delivered to the antenna input, and W_a is the energy available from the source. In the case of a receive antenna, W_d is the energy delivered to the load attached to the antenna, and W_a is the energy available at the antenna terminals. By reciprocity, the transmit and receive matching efficiencies are identical for any antenna made of linear, reciprocal materials. Lamensdorf [19] gives matching efficiency as

$$e_m = \frac{\int_{-\infty}^{+\infty} |V^+(\omega)|^2 \left(1 - |\Gamma(\omega)|^2\right) d\omega}{\int_{-\infty}^{+\infty} |V^+(\omega)|^2 d\omega} \quad (4.39)$$

where $V^+(\omega)$ is the incident voltage, and $\Gamma(\omega)$ is the reflection coefficient. This value of Γ depends on the value of normalizing characteristic impedance chosen, as does V^+. An alternative and equivalent expression that may be useful is

$$e_m = \frac{\int_{-\infty}^{+\infty} |V_g(\omega)|^2 \frac{\text{Re}(Z_A(\omega))}{|Z_g(\omega) + Z_A(\omega)|^2} d\omega}{\int_{-\infty}^{+\infty} \frac{|V_g(\omega)|^2}{4\,\text{Re}(Z_g(\omega))} d\omega} \quad (4.40)$$

where V_g is the open-circuit source voltage, Z_A is antenna impedance, and Z_g is the source impedance.

It is not possible, even in principle, to obtain a perfect impedance match between source and antenna, except for sinusoidal steady-state inputs. For a finite-time pulse of any shape, it can be shown that it is never possible to couple 100 percent of the available source energy into a finite bandwidth antenna, even with an infinitely complex matching network. This is because (1) a time-limited pulse cannot be perfectly band limited in frequency, and (2) even within a limited band, an antenna can only be perfectly matched at a limited number of discrete points. However, for sufficiently large antennas ($ka > 1$ at the lower band edge) and pulses with most of their energy above this lower band edge, it is relatively straightforward to obtain an antenna that will have very high matching efficiency, and with care in the design, it is possible to go substantially smaller.

4.3.1.2 Waveform-Independent Matching Efficiency

The matching efficiency described above is a function of the particular source waveform employed. To get a simpler expression that is a function of the antenna alone, one can note that UWB regulatory constraints effectively favor the use of waveforms with significant spectral content distributed across the entire band of interest and little content outside. If one assumes a spectrum for V^+ that is constant in magnitude across the band $\omega_L < \omega < \omega_H$ and zero elsewhere, one obtains

$$e_m = \frac{1}{\omega_H - \omega_L} \int_{\omega_L}^{\omega_H} \left(1 - |\Gamma(\omega)|^2\right) d\omega \qquad (4.41)$$

or, alternatively, with a constant spectrum for V_g, one obtains

$$e_m = \frac{\int_{\omega_L}^{\omega_H} \frac{\text{Re}(Z_A(\omega))}{|Z_g(\omega) + Z_A(\omega)|^2} d\omega}{\int_{\omega_L}^{\omega_H} \frac{1}{4\text{Re}(Z_g(\omega))} d\omega} \qquad (4.42)$$

4.3.1.3 Estimates of the Matching Efficiency

Given a reflection coefficient that varies over the range $\Gamma_{min} < |\Gamma| < \Gamma_{max}$, the matching efficiency is subject to the bounds

$$1 - \Gamma_{max}^2 < e_m < 1 - \Gamma_{min}^2 \qquad (4.43)$$

regardless of waveform. Specifying a minimum 10 dB return loss (1.92:1 VSWR) guarantees a matching efficiency of at least 90 percent.

4.3.1.4 Definition of Energy Radiation Efficiency

The energy radiation efficiency e_r is a measure of how much of the energy delivered to the antenna input is converted into radiated energy:

$$e_r = \frac{W_r}{W_d} \tag{4.44}$$

where W_r is the total radiated energy due to a pulse, and W_d is the energy delivered to the antenna input. For a receive antenna,

$$e_r = \frac{W_a}{W_c} \tag{4.45}$$

where W_a is the available pulse energy at the antenna terminals, and W_c is the energy from an arriving pulse field that is captured by the antenna. As in the case of matching efficiency, the receive and transmit radiation efficiencies are identical for linear reciprocal antennas.

Given the radiated electric fields $E(R,\theta,\phi,t)$ of a transmitting antenna, the radiated energy can be computed in the time or frequency domain as

$$\begin{aligned}W_r &= \lim_{R\to\infty} \int_{-\infty}^{\infty}\int_0^{2\pi}\int_0^{\pi}\frac{1}{\eta_0}|\vec{E}(R,\theta,\phi;t)|^2 R^2 \sin\theta\, d\theta\, d\phi\, dt \\ &= \lim_{R\to\infty}\frac{1}{2\pi}\int_{-\infty}^{\infty}\int_0^{2\pi}\int_0^{\pi}\frac{1}{\eta_0}|\vec{E}(R,\theta,\phi;\omega)|^2 R^2 \sin\theta\, d\theta\, d\phi\, d\omega\end{aligned} \tag{4.46}$$

The energy delivered to the antenna input can be computed in the frequency domain from the open-circuit source voltage V_g, the terminal voltage V_{in}, or the incident voltage V^+:

$$\begin{aligned}W_d &= \frac{1}{2\pi}\int_{-\infty}^{+\infty}|V_g(\omega)|^2 \frac{\mathrm{Re}(Z_A(\omega))}{|Z_g(\omega)+Z_A(\omega)|^2}d\omega \\ &= \frac{1}{2\pi}\int_{-\infty}^{+\infty}|V_{in}(\omega)|^2 \frac{\mathrm{Re}(Z_A(\omega))}{|Z_A(\omega)|^2}d\omega \\ &= \frac{1}{2\pi Z_0}\int_{-\infty}^{+\infty}|V^+(\omega)|^2(1-|\Gamma(\omega)|^2)d\omega\end{aligned} \tag{4.47}$$

In the time domain, it can be computed from the incident and reflected voltages:

$$W_d = \frac{1}{Z_0}\int_{-\infty}^{\infty}(V^+(t))^2 - (V^-(t))^2\, dt \tag{4.48}$$

The energy radiation efficiency is due to ohmic losses in the antenna materials or intentionally added resistive loading. Small, resonant structures, such as folded antennas, are particularly susceptible to ohmic loss due to circulating reactive energy. Even good metallic conductors and low loss-tangent dielectrics can have significant loss near resonance frequencies.

4.3.1.5 Definition of Energy Efficiency

The energy efficiency e_w, in the case of a transmit antenna, is the ratio of radiated energy to available source energy. In the case of a receive antenna, it is the energy delivered to the load over the energy captured from the incident field by the antenna. In both cases, it is the product of the radiation efficiency and the matching efficiency. For the transmit case,

$$e_w = \frac{W_{rad}}{W_a} = \frac{W_{rad}}{W_d}\frac{W_d}{W_a} = e_r e_m \qquad (4.49)$$

By combining the expressions given in the previous sections, a number of different, but equivalent, formulas for the energy efficiency can be given; for example,

$$e_w = \frac{\lim_{R\to\infty} \int_0^{2\pi}\int_0^{\pi}\int_{-\infty}^{+\infty} |\vec{E}(R,\theta,\phi;\omega)|^2 R^2 \sin\theta \, d\omega \, d\theta \, d\phi}{\eta_0 \int_{-\infty}^{+\infty} |V_g^2(\omega)|^2 \frac{1}{4\mathrm{Re}(Z_g(\omega))} d\omega} \qquad (4.50)$$

in the frequency domain. For the special case where the source impedance is real and constant with frequency, an expression in the time domain is

$$e_w = \frac{\lim_{R\to\infty} \int_0^{2\pi}\int_0^{\pi}\int_{-\infty}^{+\infty} |\vec{E}(R,\theta,\phi,t)|^2 R^2 \sin\theta \, dt \, d\theta \, d\phi}{\eta_0 \int_{-\infty}^{+\infty} V_g^2(t)/(4R_g) dt} \qquad (4.51)$$

4.3.1.6 Correlated Energy Efficiency

In applications where it is desired for the radiated field to have a particular waveform in the time domain, it is useful to have a measure of how well an antenna transfers source energy into the desired waveform. We can define an arbitrary template function with the separable form

$$\vec{T}(t,\theta,\phi) = T(t)\vec{a}(\theta,\phi) \qquad (4.52)$$

where $T(t)$ is the desired waveform, presumably not a function of direction, and $\vec{a}(\theta,\phi)$ is the desired polarization pattern. Then, we define correlated energy efficiency as

$$e_C = \frac{\int_0^{2\pi}\int_0^{\pi} \max_\tau \left[\int_{-\infty}^{+\infty} \vec{E}(R,\theta,\phi,t)\cdot\vec{a}(\theta,\phi)T(t-\tau)Rdt \right]^2 \sin\theta\, d\theta\, d\phi}{\eta_0 \left[\int_{-\infty}^{+\infty} |\vec{T}(t)|^2 dt \right] \left[\int_{-\infty}^{+\infty} V_g^2(t)/(4R_g)dt \right]} \quad (4.53)$$

in the case where the source impedance is purely real. The correlated energy efficiency is the fraction of available source energy that is radiated in the correct polarization with the correct waveform. It is, of course, dependent on the shape (but not amplitude) of the input waveform. The maximization with respect to τ is necessary to take into account possible variations in time delay as the direction varies.

4.3.2 Energy Pattern, Energy Gain, and Energy Effective Area

4.3.2.1 Energy Pattern

The energy pattern [19] is the radiated energy per unit solid angle as a function of direction:

$$U_E(\theta,\phi) = \frac{1}{\eta_0} \int_{-\infty}^{+\infty} |\vec{E}_{rad}(R,\theta,\phi,t)| R^2 \, dt \quad (4.54)$$

It has units of joules/steradian. It is dependent on both the shape and amplitude of the input waveform and is typically defined for a single isolated input pulse. An equivalent frequency-domain expression is

$$U_E(\theta,\phi) = \frac{1}{2\pi}\frac{1}{\eta_0} \int_{-\infty}^{+\infty} |\vec{E}(R,\theta,\phi,\omega)| R^2 \, d\omega \quad (4.55)$$

The energy pattern can also be expressed in terms of the vector effective length $\vec{h}$:

$$U_E(\theta,\phi) = \frac{1}{2\pi}\frac{1}{\eta_0} \int_{-\infty}^{+\infty} |\vec{h}_e(\theta,\phi,\omega)|^2 \frac{|V_g(\omega)|^2}{|R_g + Z_A(\omega)|^2} \beta^2 \, d\omega \quad (4.56)$$

The energy pattern as defined only applies to a transmitting antenna; the usual equivalence of transmitting and receiving pattern and gain for narrowband antennas does not apply, except under certain conditions, discussed in Section 4.3.2.3.

4.3.2.2 Energy Gain

The energy gain gives the energy radiated per unit solid angle in a given direction, relative to that radiated by an isotropic, lossless, perfectly matched radiator driven by a source with the same available energy and same pulse shape. It can be derived from the energy pattern as

$$G_E(\theta,\phi) = e_W \frac{4\pi U_E(\theta,\phi)}{\int_0^{2\pi}\int_0^{\pi} U_E(\theta,\phi)\sin\theta\, d\theta\, d\phi} \tag{4.57}$$

where e_W is the energy efficiency defined in Section 4.3.1.5. It is dependent on the shape, but not the absolute amplitude, of the input pulse.

The energy gain can also be derived in terms of the conventional frequency-domain gain:

$$G_E(\theta,\phi) = \frac{\int_{-\infty}^{+\infty} |V_A(\omega)|^2 \operatorname{Re}(1/Z_A(\omega)) G(\theta,\phi,\omega)\, d\omega}{\int_{-\infty}^{+\infty} |V_S(\omega)|^2 / 4\operatorname{Re}(Z_g(\omega))\, d\omega} \tag{4.58}$$

where V_A is the voltage at the antenna terminals, and G is the gain. In the special case of an antenna matched to the source at all frequencies, the result is

$$G_E(\theta,\phi) = \frac{\int_{-\infty}^{+\infty} |V_S(\omega)|^2 G(\theta,\phi,\omega)\, d\omega}{\int_{-\infty}^{+\infty} |V_S(\omega)|^2\, d\omega} \tag{4.59}$$

which is simply the gain weighted by the spectrum of the input pulse.

4.3.2.3 Energy Effective Area

The energy gain and pattern as defined above apply only to transmitting antennas since the transmit and receive energy gains are, in general, different. One can define a received energy pattern U_R as the energy received due to an incident pulsed plane wave from a given angle:

$$U_R(\theta,\phi) = \frac{1}{2\pi} \int_{-\infty}^{+\infty} \left| \frac{\vec{E}^{inc}(\omega)\cdot\vec{h}(\theta,\phi,\omega)}{Z_A(\omega)+Z_L(\omega)} \right|^2 R_L(\omega)\, d\omega \tag{4.60}$$

where $\vec{h}$ is the effective height, $\vec{E}^{\text{inc}}(\omega)$ is the Fourier transform of the incident electric field measured at the origin, and $R_L(\omega)$ is the real part of the load impedance. It is dependent on the waveform shape and amplitude. It is not, in general, equivalent to the expression for U_E, the transmit energy pattern, given in Section 4.3.2.1, unless very specific conditions are met. For example, if all of the following conditions are met, then the receive and transmit energy patterns will be equal to within a constant of proportionality:

- The impedance is matched in both cases.
- The incident field in the receive case is copolarized.
- The incident field waveform is proportional to the time derivative of the source voltage.

The *energy effective area* can be defined analogously to the conventional effective area. It is the energy delivered to the load divided by the energy per unit area in the incoming wave:

$$A_R(\theta,\phi) = \frac{\eta_0 \int_{-\infty}^{+\infty} \left|\frac{\vec{E}^{\text{inc}}(\omega)\cdot\vec{h}(\theta,\phi,\omega)}{Z_A(\omega)+Z_L(\omega)}\right|^2 R_L(\omega)d\omega}{\int_{-\infty}^{+\infty}\left|\vec{E}^{\text{inc}}(\omega)\right|^2 d\omega} \qquad (4.61)$$

4.3.3 Measures of Fidelity, Dispersion, and Time Spreading

4.3.3.1 Introduction

It should be pointed out that "pulse distortion" in the context used here and in many other antenna references is not at all similar to nonlinear distortion, such as harmonic or intermodulation distortion. Rather, it is a linear phenomenon and intrinsically causes no loss of information. On the other hand, its effects can be significant when finite time gating and sampling rates are involved.

It can be quantitatively stated that the accurate transmission of a time-domain pulse requires the satisfaction of the distortionless transmission criteria in which the transfer function has the form

$$H(\omega) = Ae^{-jKf} \qquad (4.62)$$

where A and K are constants. It is clear that the inverse Fourier transform of the frequency-domain representation of a signal multiplied by this transfer function gives a time-delayed replica of the time-domain signal. Thus, any deviation from this ideal transfer function can be considered pulse distortion or dispersion. This deviation from the ideal can result from nonlinear variation of phase with frequency or, equivalently, frequency-dependent group delay. Of course, the variation of transfer function magnitude with frequency also contributes to pulse distortion. Moreover, for a causal transfer function, the magnitude and phase of the transfer function are related by the Bode relationship.

The term *dispersion* is traditionally associated with the spreading of time-domain pulses, often due to frequency dependence of material parameters such as permittivity. We note here that any failure of a linear system to satisfy the distortionless transmission criteria will result in the distortion of time-domain pulses. There are numerous different mechanisms that can spread or compress a pulse. Some of these are allpass phenomena that only effect the phase of the transfer function, while others affect both the magnitude and the phase. In Section 4.5.9, we examine two of these, ringing and chirping.

As will be seen, antennas are a major contributor to the distortion of time-domain pulses. However, the use of equalization can, at least in static cases, compensate for pulse distortion. It is often noted that when pulse dispersion is so great that successive transmitted pulses overlap in time, unrecoverable loss of information has occurred. While extreme dispersion can distort pulses and symbols to the point that they are perhaps unrecognizable to one examining them with a high-speed oscilloscope, so long as the system is linear, the output for each unique input is unique. Thus, the dispersion cannot, in the absence of noise, cause of a loss of information. However, the amount of signal processing required to recover the signal and the time required to sample the signal sufficiently might be prohibitive.

4.3.3.2 Fidelity

Lamensdorf and Susman [19] defined a distortion parameter between two scalar time signals, $r(t)$ and $f(t)$, by means of the autocorrelation of the difference between the radiated electric field and an appropriate template function:

$$d = \min_{\tau} \int_{-\infty}^{\infty} \left| \hat{r}(t+\tau) - \hat{f}(t) \right|^2 dt$$
$$= \min_{\tau} \int_{-\infty}^{\infty} \left[\hat{r}(t+\tau)^2 + \hat{f}(t)^2 - 2\hat{r}(t+\tau)\hat{f}(t) \right] dt \quad (4.63)$$

where

$$\hat{r}(t) = \frac{r(t)}{\left[\int_{-\infty}^{\infty} |r(t)|^2 dt \right]^{1/2}}$$

and

$$\hat{f}(t) = \frac{f(t)}{\left[\int_{-\infty}^{\infty}|f(t)|^2\,dt\right]^{1/2}}$$

It is appropriate to minimize the quantity with respect to an expected time delay. The signals could be a scalar component of the radiated field, the received voltage, or numerous other system quantities. When the signal being scrutinized is radiated electric field, it is not immediately obvious to what the signal should be compared since distortion upon transmission is expected.

The distortion parameter can be related to a fidelity parameter, F, which is a cross correlation of the radiated electric field with a template function.

$$d = \min_{\tau}\left\{\int_{-\infty}^{\infty}[\hat{r}(t+\tau)]^2\,dt + \int_{-\infty}^{\infty}[\hat{f}(t)]^2\,dt - 2\int_{-\infty}^{\infty}\hat{f}(t)\hat{r}(t+\tau)\,dt\right\}$$

$$d = \min_{\tau} 2\left[1 - \underbrace{\int_{-\infty}^{\infty}\hat{f}(t)\hat{r}(t+\tau)\,dt}_{\text{cross correlation}}\right] = \min_{\tau}[2 - 2F] \quad (4.64)$$

$$F = \max_{\tau}\underbrace{\int_{-\infty}^{\infty}\hat{f}(t)\hat{r}(t+\tau)\,dt}_{\text{cross correlation}}$$

Note that the cross correlation of the two functions is maximized with respect to the time delay. As noted in [19], the choice of this template function is rather arbitrary. The anticipated transmitted field is most likely a better choice for the template than the shape of the input signal. The fidelity or, equivalently, distortion characteristics of an antenna are contained in the correlation coefficient pattern as described in [29].

4.3.3.3 Flatness Function

As previously described in Section 4.2.2.2, the radiated electric field is related to the open-circuit source voltage by

$$\vec{E}^{\text{rad}}(\theta,\phi,R,\omega) = -V_g(\omega)\left[j\eta_0 \frac{1}{R_g + Z_A(\omega)}\beta\vec{h}_e(\theta,\phi,\omega)\right]\frac{e^{-j\beta R}}{4\pi R} \quad (4.65)$$

The quantity inside the brackets is dimensionless and effectively describes the frequency response of a transmit antenna, including both mismatch and variation in the effective length with frequency. We will call this quantity the *flatness*, given by $K(\omega,\theta,\phi)$.

$$K(\theta,\phi,\omega) = j\eta_0 \frac{1}{R_g + Z_A(\omega)} \beta \vec{h}_e(\theta,\phi,\omega) \qquad (4.66)$$

4.3.3.4 Measures of Time Spreading

In certain applications, for example, pulse radio without equalization or other protection against intersymbol interference, it is desirable for the radiated energy to remain compact in time. In particular, long-duration ringing due to narrowband resonances in the antenna response may be troublesome. A measure of how well an antenna keeps the energy within a particular time slot is the time-windowed energy efficiency e_{WT}. This is identical to the energy efficiency defined in Section 4.3.1.5 above, except that the time limits are restricted to a finite interval of width T. For example, in the case of constant source resistance, e_{WT} can be expressed in the time domain as

$$e_{WT} = \frac{\max_{\tau} \int_0^{2\pi}\int_0^{\pi}\int_{\tau}^{\tau+T} \lim_{R\to\infty} |\vec{E}(R,\theta,\phi,t)|^2 R^2 \sin\theta \, dt \, d\theta \, d\phi}{\eta_0 \int_{-\infty}^{+\infty} V_g^2(t)/(4R_g) \, dt} \qquad (4.67)$$

Since e_{WT} compares radiated energy in the time window to available source energy, it takes into account mismatch and ohmic loss, as well as ringing and time spreading of the pulse. To isolate the effects of ringing and time spreading alone, one can find the fraction e_T of the total radiated energy that is in the time window:

$$e_T = \frac{\max_{\tau} \int_0^{2\pi}\int_0^{\pi}\int_{\tau}^{\tau+T} \lim_{R\to\infty} |\vec{E}(R,\theta,\phi,t)|^2 R^2 \sin\theta \, dt \, d\theta \, d\phi}{\int_0^{2\pi}\int_0^{\pi}\int_{-\infty}^{+\infty} \lim_{R\to\infty} |\vec{E}(R,\theta,\phi,t)|^2 R^2 \sin\theta \, dt \, d\theta \, d\phi} \qquad (4.68)$$

An alternative is to specify the time window T such that e_T reaches some specified percentage, for example, 99 percent power. One could thus define T_{99} as

$$T_{99} = \{T \mid e_T = 0.99\} \qquad (4.69)$$

The parameters e_{WT}, e_T, and T_{99} as we have defined them above are averaged over the entire pattern. By omitting the integrations with respect to angle, one could specify a directional e_T or T_{99} as a function of θ and ϕ:

$$e_T(\theta,\phi) = \frac{\max_{\tau} \int_{\tau}^{\tau+T} \lim_{R\to\infty} |\vec{E}(R,\theta,\phi,t)|^2 R^2 \sin\theta\, dt}{\int_{-\infty}^{+\infty} \lim_{R\to\infty} |\vec{E}(R,\theta,\phi,t)|^2 R^2 \sin\theta\, dt} \qquad (4.70)$$

A directional e_{WT} can be defined by omitting the integrations with respect to angle and normalizing to the energy per steradian that would be radiated by an isotropic, lossless matched radiator:

$$e_{WT}(\theta,\phi) = \frac{\max_{\tau} \int_{\tau}^{\tau+T} \lim_{R\to\infty} |\vec{E}(R,\theta,\phi,t)|^2 R^2 \sin\theta\, dt\, d\theta\, d\phi}{\frac{1}{4\pi}\eta_0 \int_{-\infty}^{+\infty} V_g^2(t)/(4R_g)\, dt} \qquad (4.71)$$

4.3.4 Measures of Correlated Energy

4.3.4.1 Correlated Energy Pattern

In Section 4.3.1.6, we mentioned that in some UWB applications, it is desirable to have the antenna radiate some particular waveform $T(t)$ with a particular polarization $\vec{a}(\theta,\phi)$. The correlated energy pattern U_C is the energy per unit solid angle as a function of direction, with only the energy radiated in the correct polarization and desired waveform counted [28]:

$$U_C(\theta,\phi) = \max_{\tau} \frac{\left[\int_{-\infty}^{+\infty}(\vec{E}(R,\theta,\phi,t)\cdot\vec{a}(\theta,\phi)T(t-\tau))R\, dt\right]^2}{\eta_0 \int_{-\infty}^{+\infty}|\vec{T}(t)|^2\, dt} \qquad (4.72)$$

It depends on both the input waveform shape and amplitude. As in the case of correlated energy efficiency, the maximization with respect to t is necessary to take into account possible variations in the time delay with respect to direction.

4.3.4.2 Correlated Energy Gain

The correlated energy gain G_C is the ratio of energy per unit solid angle, radiated in the correct polarization $\vec{a}(\theta,\phi)$ and desired waveform $T(t)$, to the energy that would be radiated by a "perfect" isotropic antenna with the same available source energy.

The perfect isotropic antenna used for comparison would radiate all the energy into the correct polarization with the desired waveform and would be perfectly impedance matched. An expression for G_C is

$$G_C(\theta,\phi) = \frac{4\pi e_C U_C(\theta,\phi)}{\int_0^{2\pi}\int_0^{\pi} U_C(\theta,\phi)\sin\theta d\theta d\varphi} \tag{4.73}$$

where e_C is the correlated energy efficiency defined in Section 4.3.1.6. G_C does depend on the input waveform shape, but it does not depend on its amplitude.

4.3.4.3 Correlation Pattern

If only the distortion of the radiated waveform, and not its amplitude, is of interest, then the correlation pattern is useful:

$$\rho(\theta,\phi) = \max_{\tau}\left[\frac{\int_{-\infty}^{+\infty}(\vec{E}(R,\theta,\phi,t)\cdot\vec{a}(\theta,\phi)T(t-\tau))Rdt}{\sqrt{\int_{-\infty}^{+\infty}|T(t)|^2 dt}\sqrt{\int_{-\infty}^{+\infty}|\vec{E}(R,\theta,\phi,t)|^2 R^2 dt}}\right]^2 \tag{4.74}$$

If the polarization and time-domain shape of the radiated waveform are correct, the correlation will be one; if the radiated waveform is orthogonal to the desired template, the correlation will be zero.

4.3.4.4 Choice of Template

As pointed out in [19], the choice of a template function for the computation of a fidelity pattern is somewhat arbitrary; that is, it is not clear what the radiated field should be compared to in order to assess fidelity. The same is true for related parameters such as correlated gain or correlation coefficient. The choice of $T(t)$ for any of the correlated quantities discussed above will depend on the application; however, if one is mainly interested in the variation of pulse shape with angle, rather than in obtaining a particular shape, it is useful to use the waveform in the bore site or nominal direction of maximum radiation. With the goal of compensating for distortion in mind, the template should, perhaps, be chosen to be the transmitted field in a particular reference direction; that is, it might be reasonable to assume that the antenna is to be operated with an equalizer that compensates for signal distortion in the direction of maximum radiation. This is the template selected in most of the numerical examples given in later sections. For a dipole or biconical dipole, it is simply the shape of the radiated electric field in the H-plane of the antenna.

4.3.4.5 Interaction of Impedance Match with Energy and Correlation Patterns

In conventional narrowband antenna work, it is common to think of the radiation pattern and the impedance match as completely decoupled problems. This is not the case when dealing with correlated energy and correlation patterns. Even in the case of perfect power transfer with an allpass phase-shift function, the correlated energy and correlation coefficient patterns defined here can be affected by the dispersion; that is, the variation of insertion phase with angle affects how well signals transmitted in different directions will be correlated with one another. Thus, it is possible that, when the traditional (power) radiation pattern of the antenna is frequency dependent, the action of matching components placed at the input of an antenna can change its correlated energy pattern and correlation coefficient pattern.

The biconical antenna serves as a practical and interesting example of the interplay between the antenna descriptors [30]. The biconical dipole antenna with a 60° cone angle is actually a reasonably good antenna for 3.1–10.6 GHz UWB applications in that it can exhibit about two octaves of impedance bandwidth with nominally 2:1 VSWR using 200Ω source impedance.

The matching efficiency as defined in Section 4.3.1.1 is embedded in the energy gain of an antenna. Table 4.2 gives the matching efficiency for the 60° biconical antenna for a normalized pulse width of $0.058 \times (\text{length}/c_0)$ and four different source impedances.

It can be seen that the matching efficiency increases with the source impedance. The physical explanation is that, while near the fundamental series resonance, the real part of the impedance is near 50Ω, above the fundamental series resonance, the magnitude of the input impedance is relatively high over the next several octaves. Figures 4.7 and 4.8 show the energy gain and fidelity patterns of the antenna for the same source impedances and pulse width. As can be seen, the shape of the patterns depends on the source impedance. While more energy is coupled into the antenna for higher source impedances, this energy is concentrated in the frequency range over which the power patterns of the antenna vary rapidly with frequency. Thus, while increasing the source impedance improves the matching efficiency, it degrades the fidelity pattern.

$R_S = 50$	$\eta = .436$
$R_S = 100$	$\eta = .645$
$R_S = 200$	$\eta = .813$
$R_S = 400$	$\eta = .832$

Table 4.2 Matching Efficiency for 60 degree Biconical Antenna with Gaussian Pulse Excitation; $s = .058 \times (L/c_0)$. From [89].

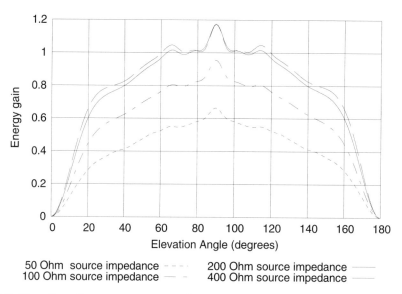

Figure 4.7 The effect of source resistance on the energy gain of 60 degree biconical antenna with Gaussian pulse excitation. The natural width of the Gaussian pulse is .058*(length/c_0). From [29].

Except under a very restricted set of conditions, the time-domain pulse waveform of the radiated field from a UWB antenna will vary with direction.

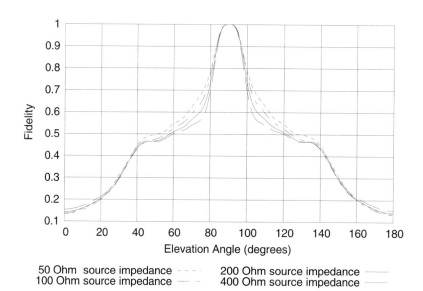

Figure 4.8 The effect of source resistance on the correlation coefficient of 60 degree biconical antenna with Gaussian pulse excitation. The natural width of the Gaussian pulse is .058*(length/c_0). From [29].

4.3.4.6 Conditions for Direction-Independent Distortion

Except under a very restricted set of conditions, the time-domain pulse waveform of the radiated field from a UWB antenna will vary with direction. This can be seen clearly in the results shown in Figure 4.5. This effect means that any equalization or compensation scheme to restore the waveform shape must be adaptive in a mobile system.

For the radiated waveform shape to be independent of direction, which we will call *direction-independent distortion* (DID), it is necessary that particular conditions be met. It is not sufficient for the frequency-domain power pattern to be independent of frequency; in fact, even an isotropic antenna can have a direction-dependent waveform. The phase pattern must also satisfy particular constraints. This restriction impacts frequency-independent antennas, which often have a stable power pattern but a moving phase center as the frequency is changed.

For an antenna to be DID, the following conditions must be met. First, the conventional frequency-domain gain G must be separable into a function of angle and a function of frequency:

$$G(\theta,\phi,\omega) = g(\theta,\phi)f(\omega) \tag{4.75}$$

Second, the phase pattern Ψ must be a function of angle only multiplied by ω, except for phase changes that are a multiple of π:

$$\Psi(\theta,\phi,\omega) = \omega \tau_d(\theta,\phi) + m\pi \tag{4.76}$$

Jumps of phase by an amount equal to π can occur in the phase pattern (for example, moving from a main lobe into an adjacent side lobe). In practice, no antenna will exactly meet these criteria, but they can be approximated. Any antenna will tend to meet the criteria as it becomes very small electrically; however, then restrictions on impedance bandwidth, as discussed in Section 4.1.1, will dominate the performance.

4.4 Radiation-Pattern Control, Balance, and Ground-Plane Independence

This section addresses several ancillary topics, including the control of radiation patterns over the intended frequency range, balance, and the interaction of the antenna with its platform. None of these topics is unique to the UWB area; however, the broad bandwidth requirement, as well as some anticipated mobile applications, bring some of these topics to the forefront.

4.4.1 Radiation-Pattern Control

The importance of an antenna's radiation-pattern characteristics in a communications link is strongly tied to the environment in which the antenna must operate. For example, in a fixed link, maximum gain in a particular direction is generally desirable, as is side-lobe suppression in other directions. When an antenna is to be operated on a mobile platform, this is clearly not true. Many mobile applications call for an "isotropic" antenna. It should be clarified that a true isotropic one-port antenna cannot exist. There are so-called isotropic field probes consisting of three orthogonal dipoles. However, these are actually combinations of antennas and detectors. It can be shown that the polarization state of the far fields of an antenna with an isotropic power pattern is elliptical and, moreover, must take on every possible value of axial ratio as the spherical angles are varied [31]. From the analysis presented in [31] and the previous discussion on direction-independent distortion, we can conclude that it is not possible to make an isotropic antenna with direction-independent distortion. The important point here is that when one considers a line-of-sight (LOS) link with one or both antennas mobile, it is not possible to specify that there will never be polarization mismatch when single-port antennas are employed. Moreover, the direction-dependent distortion of such an antenna will make equalization more complicated.

The radiation pattern of an electrically small antenna is typically invariant with frequency so long as the antenna remains electrically small; that is, the pattern for a linear or tapered dipole will remain dipolelike as long as the antenna is electrically small. Unfortunately, this is also the frequency range in which effective power transfer is difficult to maintain over a broad bandwidth. Thus, the control of radiation-pattern features becomes important if the antenna is to be operated at frequencies for which it is not electrically small. The problem of designing a useful UWB single-element antenna can often be simplified to impedance matching at the lower end of the operating frequency range and pattern control at the high end. When one considers the problem of maintaining a pattern, such as that of the TM_{01} spherical mode over a broad frequency range, it becomes clear that there is no physical limit concerning the high-frequency operation of such an antenna; that is, it should be possible to maintain the pattern in a highpass sense. However, close examination of the patterns of many of the published monopole antennas that approach the fundamental limits for performance in the frequency range in which they are electrically small shows that the patterns are very complex at higher frequencies. This is true for the famous Goubau antenna [32] and similar multielement monopoles. Features that provided compensation at low frequencies, such as folding, radiate significantly at higher frequencies. It is very difficult indeed to design an efficient, well-matched, electrically small antenna that maintains a monopole or dipole pattern in the frequency range for which it is of moderate electrical size.

4.4.2 Balanced versus Unbalanced Antennas

Antennas presenting a plane of symmetry and a two-conductor feed are occasionally referred to as balanced antennas. However, in the presence of a ground or other large counterpoise, such an antenna possesses both a differential and a common-mode radiating structure. This concept is important because, under some conditions, the effects of the common-mode structure can dominate those of the differential structure; that is, the common-mode structure, being generally at least twice as large as the differential structure, may actually be more apt to radiate than would the differential structure. It is certainly true that a linear monopole or any monopole antenna is unbalanced in the sense that current must exist on the ground plane or counterpoise in order for the antenna to function; however, a geometrically symmetrical dipole antenna is not necessarily balanced in the sense that its pattern is symmetrical if there is an external unbalancing influence such as a nearby object. It must be said that there really is no device that, when placed between the input of an antenna and a transmission line, can enforce balance. That being said, the proper use of a balun might, in part, correct for asymmetrical influence and provide some degree of ground-plane independence.

4.4.2.1 Ground-Plane Independence

It is often desirable for an antenna to be able to function independently from the platform to which it is attached. With so-called ground-plane independence, the antenna can be employed, in principle, with many different devices. This would be particularly desirable for an antenna sold as a component to be attached to an existing board. Moreover, such an antenna would be less likely to cause electromagnetic interference (EMI) problems since the radio frequency currents would be more limited to the antenna structure proper.

Perhaps the simplest example of this concept is the comparison between the quarter-wave monopole and the end-fed half-wave dipole employing a quarter-wave sleeve choke integral to one-half of the dipole. The quarter-wave monopole definitely exploits the device to which it is attached as a counterpoise and, thus, is approximately half as long as the dipole. On the other hand, the performance of such an antenna is strongly dependent on the characteristics of the radio it is attached to. For handheld transceivers, the radio frequency currents generated on the radio by a monopole also couple strongly to the user, greatly increasing loss, as well as specific absorption rate (SAR). For this reason, such antennas are almost never employed in high-power applications on handheld units.

On the other hand, it can be seen that the half-wave dipole is essentially complete in itself since the integral choke sleeve effectively places an open circuit between the dipole and the radio. However, a simple quarter-wave choke is inherently narrowband; thus, it would be a mistake to state that in the UWB case the antenna operation is truly independent of the device to which it is attached. Chokes with broader operating frequency ranges (such as helical chokes) have been developed, but even in this case, the quasistatic near fields of a linear dipole have a dominant radial electric field component that will couple strongly to almost anything placed near the end of the dipole.

The end-driven dipole can be used to illustrate the problem of ground-plane independence further. Two quarter-wave chokes placed in cascade have been found to ameliorate the problem significantly, while coupling to the exterior of the second choke is minimal as it is well below fundamental half-wave resonance. Of course, this is a narrowband approach. However, broadband extensions of this approach, such as helical chokes and a tapered dipole, are straightforward. We can glean from this that a "balanced," or complete, antenna is required, as is at least one more device to decouple the near field from the mounting platform further. These greatly increase the size of the unit.

It can be said that for any antenna that behaves as an electric dipole, the near-field coupling near either tip of the dipole is so strong as to overwhelm the action of any attempt to decouple the antenna. Thus, having an electric dipole antenna aligned such that it is normal to the conducting surface of the device to which it is mounted will almost always result in significant interaction. This, unfortunately, is the arrangement that usually produces a favorable or constructive image for an electric dipole. Of course, other types of low-gain antennas are possible, but true ground-plane independence is hard to achieve for any antenna type, especially over a broad bandwidth.

In summary, in order to implement a platform-independent antenna, one must choke off common-mode current as well as effectively eliminate near-field coupling. It should be noted that, for many common antenna designs, near-field coupling simply cannot be eliminated, making true ground-plane independence impossible.

4.4.2.2 Effects of Finite Counterpoise

Since it is likely that many consumer UWB devices will operate on small to moderate-sized platforms, a study of platform effects is worthwhile. In fact, the utility of so-called ground-plane antennas designed to operate over practically infinite ground planes is of questionable relevance.

Wheeler notes in [8] concerning fundamental limitations in antennas that, when considering the electrical size of the antenna, it is essential that all images be included. It is exactly this effect that has occasionally given rise to claims of bandwidth-size performance that appear to exceed fundamental limits. One particularly misleading geometrical arrangement is a monopole mounted on the edge of a ground plane. Such an arrangement can provide excellent performance in terms of impedance bandwidth; however, it is a mistake to take the image of the monopole as being identical to the image of the monopole that would be produced by a ground plane perpendicular to the axis of the monopole.

An analytical expression for the input impedance of a monopole mounted on the edge of an infinite ground plane has been derived [33]. The expression for the input resistance is

$$R_F = \frac{3}{8}\eta\left[C(kl) - \cot(kl)S(kl)\right].$$

Note that this expression is slightly different from that given in [33]. It would appear that equation (3) in [33] has a typographical error. The expression given here is consistent with equation (5) in [33]. When the length of the monopole is so short that the current variation on the monopole is essentially linear, the radiation resistance is given approximately as

$$R_F \approx \frac{2}{3}\eta\frac{l}{\lambda}$$

Thus, the radiation resistance varies linearly with frequency in contrast to a short monopole situated over an infinite ground plane, for which the radiation resistance varies with the square of frequency. The authors computed the input reactance using transmission-line theory and gave an analytical expression for it. While the analytical expression for input resistance agreed well with measured data, the authors found that agreement between the analytical expression for input reactance and experimental results required an additional base capacitance. Nevertheless, it is still clear that the Q of such an antenna would vary with the inverse square of the antenna size as opposed to the inverse cube, that is, if the antenna size is taken as the size of the element protruding from the edge of the ground plane. No matter how difficult the reactance might be to compute accurately, it must ultimately behave as a lumped capacitance in the asymptotic low-frequency limit. In [34], a very accurate computation of the input impedance of a monopole mounted on the edge of an infinite ground plane is computed using the exact Green's function (which is derived in [34]). Figure 4.9 gives the computed input resistance and reactance of the monopole, along with values computed from W. A. Johnson's approximate expressions. As can be seen, the radiation resistance does indeed vary linearly with frequency when the monopole is electrically short. In any case, the quasistatic capacitance behaves as a lumped element, and the Q of the antenna must therefore tend to be the inverse square of the electrical size.

However, the definition of *image* is that which would have to replace the counterpoise or ground plane to recover a free-space problem. Clearly, in the case of the monopole mounted on the edge of an infinite ground plane, the image itself is of

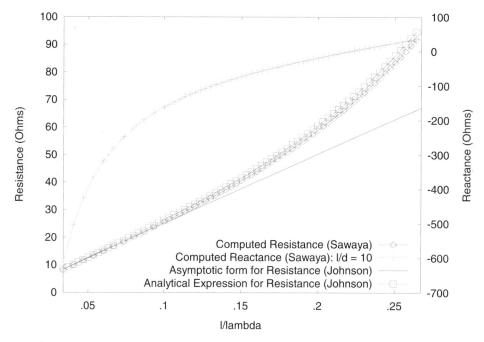

Figure 4.9 Computed input impedance components for a short monopole on the edge of a semi-infinite ground plane. The length to diameter ratio is 10.

infinite size. It has been noted that perhaps the term *monopole* is simply a misnomer for this antenna. The pattern of this antenna, while containing components similar to those of a traditional monopole, contains other components related to radiation from the edge of the ground plane. In fact, Johnson points out that the improved radiation characteristics of the antenna are due, at least in part, to radiation components other than those of traditional monopole.

4.4.3 UWB Baluns

In order to discuss baluns, it is necessary to specify exactly what is meant by balance. A balun, when properly designed, is capable of enforcing a balance in current, voltage, or some linear combination of these quantities. As noted earlier, enforcing such balance at the base of an antenna does not necessarily enforce symmetry or ground-plane independence in the antenna.

It should also be noted that the use of a separate balun for driving a UWB antenna, especially an electrically small balun, places further limitations on performance, such as loss and possibly dispersion.

Here we address a few balun types. It should be emphasized that the list is far from complete. For a very good contemporary reference on the use of baluns with pulsed antennas, the reader is referred to [35].

4.4.3.1 The Infinite Balun

While it is not strictly a balun at all, the so-called infinite balun definitely deserves mention here as it is truly the most frequency independent of all feed mechanisms and is suitable for numerous balanced antennas, including log-periodic dipole antennas (LPDAs), horns, shielded loops, spirals, conical log spirals, and undoubtedly others. This balancing mechanism is based on two principles. First, if a control volume is made small enough, the net current exiting the closed surface surrounding the volume is zero by the continuity equation. That is,

$$\oiint_{S'} \vec{J} \cdot d\vec{S} = -\frac{d}{dt} \iiint_{V'} \rho \, dV$$

where $\vec{J}$ is the current density, ρ is the volumetric charge density, and the region V' is bounded by the closed surface S'. If the control volume is made small enough, we have approximately

$$\oiint_{S'} \vec{J} \cdot d\vec{S} = 0$$

This, of course, is the basis for Kirchoff's current law. Second, for the TEM mode in any multiwire transmission line, a current balance always exists. Thus, for the TEM mode in a coaxial transmission line, the net current on the inner conductor equals the net current on the inside of the outer conductor. In short, if an antenna design allows the incorporation of an infinite balun feed structure, it should be used. Such an approach is almost always used for LPDAs.

4.4.3.2 The Marchand Balun

Of the common balun types, the most widely used design is the Marchand balun [36]. While the so-called parallel connected balun [37] is thought to provide greater bandwidth, the Marchand balun is simpler and provides adequate bandwidth for the commercial UWB frequency range. Planar versions of the Marchand balun have been developed that are compatible with planar antenna implementations. The Marchand balun is a distributed device and, hence, is not electrically small, although the use of high-permittivity materials can reduce the physical size of the device. The main point to be made here is that this simple design is actually adequate for the 3.1–10.6 GHz bandwidth anticipated for commercial UWB applications.

4.4.3.3 The Helical Common-Mode Choke

A helically wound bifilar transmission line can serve as a broadband common-mode choke and, hence, serve as a current balun. Such a common-mode choke is regularly used in differential signal paths to prevent common-mode leakage or coupling from external common-mode current caused by external fields. For applications at ultra-high frequency and below, such chokes almost always employ a magnetic core material that is usually quite lossy. Even with hexagonal ferrite materials, magnetic losses limit the applicability of ferrite materials in the microwave frequency range. In any case the common-mode choke, when loaded with magnetic materials, effectively places a resistance in the common-mode structure. If an antenna is by nature highly unbalanced, the addition of such a choke will elicit very large ohmic losses. In the case of no loading, the helical choke effectively places an inductance in the effective common-mode structure. If the common-mode structure is capacitive in nature, there is a possibility that a common-mode resonance will occur between the choke and the antenna.

4.4.3.4 Shunt-Connected Balun

The so-called coupled-line balun, Guanella balun, and shunt-connected balun all have exactly the same fundamental topology. This topology is important because, in the limiting case in which the even-mode or common-mode impedance becomes infinite, the balun becomes a true-time-delay or pulse-preserving network when the characteristic impedance of the constituent transmission lines has the optimum value, namely twice the impedance of the unbalanced port and half that of the load impedance.

4.4.3.5 Lattice Baluns

There are a number of designs based on lattice allpass phase-shift structures, including lumped-element lattice baluns as well as double-Y baluns. Note that the topology of the double-Y balun is identical to that of the lumped-element lattice balun when the Richard's transformation is applied to it. Although the lattice structure is an allpass network, many of these designs provide current balancing action only over a narrow bandwidth. Moreover, such designs, being constructed of allpass phase-shift networks, necessarily introduce dispersion into the system, unlike the shunt-connected balun. However, it is difficult to develop a balun design that is simultaneously low loss and broadband and provides current balancing action. Excellent results have been obtained using a double-Y balun with a dipole antenna for pulsed applications [35]. For a complete explanation of the interaction of the double-Y balun with the dipole, it would be necessary to consider the three-terminal equivalent network for the antenna and its feed transmission line.

4.4.3.6 180° Hybrids Employed as Baluns

It should be noted that any 180° hybrid could be employed as a balun [38]. However, such broadband hybrids often exhibit significant insertion loss. Moreover, if the hybrid is used as a 180° power divider where the sum port is terminated, it is possible that significant power will be dissipated in the sum port termination. While it might be argued that this power is related to conversion of delta mode to common mode in the antenna and, hence, could have resulted in imbalance in the absence of the hybrid action, some lack of balance is most likely a small price to pay for efficiency. In [39], it is shown that the use of a terminated hybrid or 180° power divider as a balun is equivalent to inserting a resistance into the common-mode structure. Thus, if any tendency for imbalance exists, power will be dissipated in this resistance.

In summary, three points concerning the use of baluns in UWB antenna applications should be kept in mind:

1. A balun cannot necessarily enforce antenna pattern balance. In many cases, it is difficult to enforce current balance effectively at the driving point of an antenna. Even when it is, near-field coupling can still cause the overall current distribution on a symmetrical antenna to be asymmetrical. The rigorous approach to this problem involves viewing the antenna and its platform or counterpoise as a three-terminal network. This view should be considered when the antenna is driven with a balanced transmission line or source. In some silicon-based systems, balanced 100Ω transmission lines are employed. Connecting such a system to a symmetrical antenna would greatly simplify the design and perhaps provide better performance. However, it must be remembered that an asymmetrical influence on the antenna, combined with a finite common-mode characteristic impedance of the balanced line, could lead to EMI problems.

2. Any additional network inserted between the antenna and the radio necessarily introduces loss. The smaller the electrical volume is, the greater the dissipation will be. Thus, distributed baluns will provide lower insertion loss, but at the expense of greater size.

3. If pulse fidelity is an issue, the pulse distortion intrinsic to the balun must be taken into consideration. Most coupled-line baluns exhibit a nonlinear insertion-phase variation with frequency. Such insertion phase is seldom considered since the relative phase of the balanced ports is normally the primary consideration. It is interesting to note that the canonical lattice balun actually comprises allpass phase-shift networks; thus, it is not conceivable that the through response of such a device might be exploited to compensate for dispersion in some frequency-independent antennas.

4.5 Survey of UWB Antenna Design

Most UWB antennas in the literature have not been analyzed with respect to the UWB performance parameters discussed in Section 4.3. This is due partly to UWB's only recently receiving extensive attention for communications applications and partly to the additional computational effort required. Accurate evaluation of correlated energy efficiency, for example, requires computation and storage of a full-sphere complex radiation pattern either in the time domain or at a large number of points in the frequency domain.

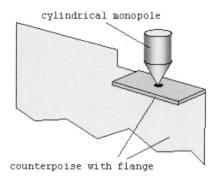

Figure 4.10 Tapered cylindrical monopole mounted near corner of 60 mm × 120 mm ground plane.

4.5.1 "Fat" Cylindrical Monopoles and Dipoles

Early in the history of wire antennas, it was recognized that increasing the wire diameter improves bandwidth due to a reduction of the stored energy in high field regions near the surface of thin wires. The biconical antenna and its monopole equivalent, discussed in Section 4.5.5, are the limiting cases of this concept. There is an extensive literature on both wire-cage and solid monopoles; we will only briefly mention some recently published examples here.

K.-L. Wong and S.-L. Chieng [40] recently presented a cylindrical monopole with a tapered input section, 20 mm in height by 10 mm diameter, mounted near the corner of a 60–120 mm vertical ground plane with flange, as shown in Figure 4.10. The antenna had return loss greater than 10 dB over a very wide bandwidth (1.84 to 10.6 GHz). The pattern, as expected, shows definite signs of radiation from currents on the electrically large counterpoise, with multiple lobes, strong azimuthal variation, and nulls greater than 10 dB even in the lower portion of the band.

Wong and Chieng [41] also presented a square cross-section monopole with an RL = 10 dB bandwidth of 1.92 to 6.63 GHz, 23 mm in height with an 8 mm × 8 mm cross-section, mounted on a conventional 150 mm × 150 mm horizontal ground plane. In contrast to the cylindrical element, a clean monopole-type pattern was obtained with less than a 2 dB azimuthal variation in gain and multiple lobes just beginning to appear at 4 GHz, one octave above the bottom band edge. The good pattern relative to the previous case can be attributed to the effects of the ground-plane arrangement.

4.5.2 Planar Monopole Antennas

The planar monopole antenna (Figure 4.11) has received a great deal of attention in the recent UWB literature due to its ease of fabrication. The basic idea is to create a "fat" structure, which reduces the Q and, thus, makes it simpler to tune the antenna at lower frequencies. At the lower frequencies, the precise shape is noncritical, which

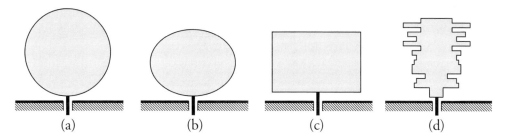

Figure 4.11 Variants of the planar monopole antenna over a ground plane. The antenna at the right is typical of planar monopoles optimized through the use of genetic algorithms (GA).

accounts for the variety of shapes that have been used with reasonably good results. When a very wide (multioctave) response is needed, the shape of the base region near the feed becomes important. In the elliptical version of planar monopole, extremely wide impedance bandwidths can be obtained, with a ratio of 8:1 (three octaves) being fairly common. Typical heights are 0.25λ to 0.35λ at the lower edge of the band.

Smaller sizes can be obtained while still covering the UWB. The shape on the right in Figure 4.11 has been optimized using a genetic algorithm [32]. It has greater than two octaves of bandwidth for 3:1 VSWR and is only 0.16λ high at the lower band edge. However, the pattern bandwidth is much smaller. At low frequency, the pattern is that of a conventional monopole, but the pattern eventually becomes that of two tapered slot antennas with their main lobes in two opposite directions.

4.5.3 Planar Monopole Antennas with a Band-Notch Characteristic

UWB systems must share their frequency allocation with existing narrowband services—for example, IEEE 802.11a systems operating from 5.15 to 5.825 GHz. These services have a much higher-power spectral density than the UWB signals, and, in many cases, band-stop filtering will be required to prevent interference. While this would normally be accomplished by a conventional filter in the radio frequency receiver front end, it is possible to design antennas with a band-notch characteristic to aid in narrowband signal rejection.

The planar monopoles described above can be modified by the addition of internal or external slot structures. A. Kerkhoff and H. Ling [32, 42] have presented square and genetic algorithm (GA) optimized designs of this type, as shown in Figure 4.12(a, b). The rejection in these designs occurs at the frequency for which the slot is approximately $\lambda/2$ in total length. High rejection (in excess of 20 dB) can be obtained at a single frequency. Yoon et al. [43] report a different arrangement, as shown in Figure 4.12(c).

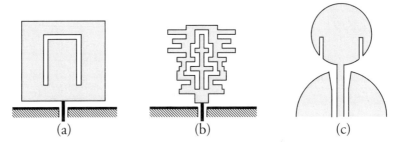

Figure 4.12 Planar monopole antennas modified to have a band-notch frequency response. In the antenna at the left [41] a half-wave interior slot is added to a standard monopole design. The antenna at the center [42] has had both the outer silhouette and the internal slot shape optimized by a genetic algorithm. The antenna at the right [43] has two exterior slots, each less than ¼ in. length.

4.5.4 Broadband Antennas Derived from the Top-Loaded Monopole

When antenna height is a primary consideration, and ease of fabrication can be sacrificed, variations on the top-loaded monopole may be considered. Top-loading of monopoles is a very old technique, originally intended solely to allow an antenna to be self-resonant at a given frequency with a reduced height. However, if modifications to the vertical element are also made, an antenna can be designed to be inherently double-tuned (or to have higher-order tuning).

One class of modifications involves the use of a combination of multiple driven and undriven vertical elements, which transforms the basic antenna impedance to a more suitable level and simultaneously creates transmission-line mode currents that can be used to double-tune the antenna structure, that is, to obtain a fourth-order frequency response. These types of antennas do not have axial symmetry and, thus, will develop asymmetry in the azimuthal pattern at the higher end of their frequency ranges.

An early example of this is the Goubau antenna [44], which has one octave of impedance bandwidth and is 0.068λ high at the lower band edge (2:1 VSWR). It has two driven and two undriven elements under a segmented top hat. The segmentation allows for inductive loops to adjust the transmission-line resonance for double tuning. It can be shown that Goubau's design has nearly optimum performance for a double-tuned, top-loaded monopole of its given aspect ratio.

Further bandwidth can be obtained with structures that effectively have higher-order tuning. L. Cobos, H. Foltz, and J. McLean [45] have recently described a modification to the Goubau antenna shown in Figures 4.13 and 4.14, with dielectric-loaded planar vertical elements, an unsegmented top hat, and an additional high-frequency vertical element that has two octaves of impedance bandwidth, as can be seen in Figure 4.15, and is 0.078λ high at the lower band edge. The dielectric loading allows the electrical length seen by the transmission-line mode to be adjusted for $\lambda/4$ resonance at a selected frequency, without the need for a segmented

UWB Antennas 119

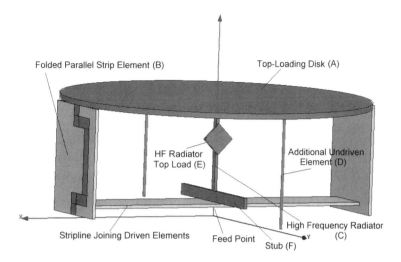

Figure 4.13 Perspective view of modified Goubau antenna. The ground plane has been omitted for clarity. The tuning stubs (F) shown at the base were not needed in the version presented here.

top disk and tuning loops. The relative widths of the strips can be chosen to control the impedance step-up level, and the thickness of the dielectric can be used to control the characteristic impedance of the transmission-line mode. A polar plot of the input reflection coefficient (see Figure 4.16) shows that the antenna is effectively eighth order.

The pattern is similar to a standard monopole at the low end of the frequency range; however, at the high end, the antenna is nearly one wavelength in diameter and has a nonuniform azimuth pattern. Nulls appear in the pattern starting at 1,200 MHz.

A second class of modifications involves the use of modified or interdigitated vertical elements. C. H. Friedman [46] reported bandwidth results matching that of Goubau with an antenna of electrically equivalent dimensions, using a single-ele-

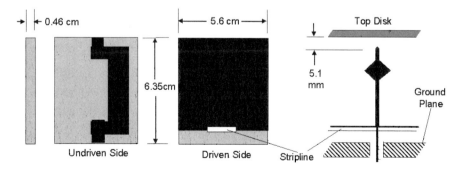

Figure 4.14 Detail of folded parallel strip element with meandering line, fabricated on FR-4 (left); detail of high frequency radiator located at center (right).

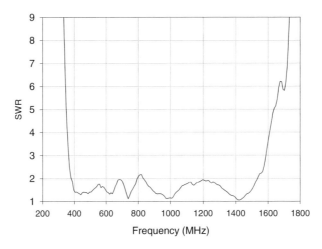

Figure 4.15 VSWR versus frequency for the modified Goubau antenna.

ment top-loaded monopole with biconical center post and a lumped-element matching network. Friedman's results, like Goubau's, are nearly optimal.

The authors have developed an interdigitated antenna with dielectric loading showing a bandwidth (2:1 SWR) extending from 2.6 to 6.0 GHz that is 18 mm high (0.16λ). More recently, an interdigitated antenna has been developed that is the same size as the Goubau antenna and covers one octave of bandwidth [47]; however, it has 3:1 VSWR over the operating band. Although the size/bandwidth/SWR trade-off achieved to date is not as good as for the multielement top-loaded anten-

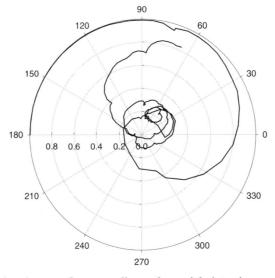

Figure 4.16 Polar plot of input reflection coefficient for modified Goubau antenna.

nas, this class of antennas has a significant advantage in that the pattern is omnidirectional at all frequencies.

4.5.5 Biconical and Bowtie Antennas

The biconical and bowtie antennas have long been known to exhibit excellent broadband performance. An excellent discussion of these is presented in [4, 48]. An experimental investigation performed by G. H. Brown and O. M. Woodward showed that two octaves of impedance bandwidth are obtainable with a biconical antenna [49]. The true biconical antenna is a conic section and, thus, can be analyzed using techniques of separable coordinates, namely, spherical coordinates. Not only does the tapering of the biconical antenna improve its impedance characteristics over those of the linear dipole, but the tapering also discriminates against higher-order spherical modes as can be shown with a spherical-mode decomposition of the far field. Many practical implementations of biconical, triangular, and other tapered antennas have been developed. Some of these concepts are shown in Figure 4.17.

The biconical dipole in its canonical form without any additional features is a reasonably good UWB antenna. With a 60° cone, it can exhibit about two octaves of impedance bandwidth with nominally 2:1 VSWR using a 200Ω source impedance. The conventional power pattern is very nearly that of the TM_{01} spherical mode over this entire range and, thus, exhibits direction-independent distortion. The pattern is more or less dipolelike up to the frequency at which the antenna is about four thirds of a wavelength in electrical length. Thus, one might expect the energy and correlated energy patterns to be very similar for pulses with spectral content limited to this frequency range. Finally, pulse reproduction is reasonably good and can be improved further at the expense of efficiency with resistive loading (see Section 4.5.7).

In many situations, other requirements, such as size, preclude the use of the biconical. A biconical dipole for the commercial UWB frequency range would be about 3.2 cm in total length. The corresponding monopole would be one-half of

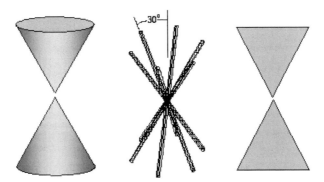

Figure 4.17 Simple biconical antenna (left); wire biconical antenna (center); bowtie antenna (right).

that but would necessarily require a large counterpoise. Nevertheless, such an antenna is invaluable for prototype development and laboratory testing.

In the EMC area, numerous versions of biconical antennas have been developed and are used extensively for both emissions and immunity testing, especially in the range of 30 to 300 MHz. The large amount of effort spent in developing these antennas has led to a few improvements in the practical design. While the frequency range requires a physically large antenna (1.4m length is typical), a wire-cage implementation is normally used instead of a solid element. When a sufficient number of wire elements are employed, the performance is similar to that of the solid element. Perhaps the principle difference is the possible coupling to interior modes.

The impedance level of the biconical antenna tends to be higher than that of a resonant half-wave dipole. Over the years, most designers have opted to employ a 4:1 impedance transforming balun, leading to an effective source impedance of 200Ω seen by the antenna. This is an improvement over a 1:1 balun (50Ω source impedance). However, Woodward's data indicates that a 2:1 impedance transformation might be closer to optimum. Biconical antennas are typically operated well below the fundamental series resonance of the antenna. In this range, the magnitude of the input impedance is large. It can be shown that, in the absence of any other matching components, the best broadband match is obtained when the source resistance is close to the magnitude of the load impedance. Thus, even though the match is poor, it is better with an effective 200Ω source impedance than with a 50Ω source impedance. The relatively high effective source impedance does, in fact, provide better pulse reproduction as the source (or load) resistance provides damping for the antenna. This is shown in Section 4.5.7. However, as with resistive loading, the trading of pulse fidelity at the expense of power or energy transfer is not a very sound approach. Thus, the use of a broadband transformer is limited.

The bowtie antenna provides similar, but slightly degraded, performance to the biconical antenna. The principal advantage of the bowtie antenna is the fact that it is flat and, thus, can be implemented in a planar medium such as printed circuit. Such cost-savings frequently far outweigh any performance degradation for massed-produced applications. Brown and Woodward have given a comparison of biconical and bowtie antennas.

In [50], the pulse performance of the biconical antenna was considered for three different pulse shapes shown in Figure 4.18. Some of these results are reproduced here to demonstrate the typical UWB performance of the biconical, as well as to give an example application of some of the UWB antenna performance parameters described in Section 4.3. In Figure 4.18, the pulses are normalized to the same peak value for convenient viewing; however, in the EM simulations which follow, they are normalized to equal source energy. The energy, correlated energy, and correlation coefficient patterns will be dependent on the pulse shape. Furthermore, the relationship between the pulse width and the antenna length is crucial. The antenna behaves fundamentally differently depending on whether the pulse is short or long compared

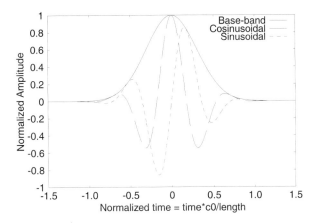

Figure 4.18 Three pulse shapes: a baseband Gaussian pulse, a cosinusoidal Gaussian burst, and a sinusoidal Gaussian burst. The period of the sinusoidal function is 1.665 times the natural width of the Gaussian function.

to the transit time L/c_0 associated with the antenna. Here L is the tip-to-tip length of the antenna.

The baseband Gaussian pulse has an open-circuit source voltage of the form

$$v(t) = e^{-(t/\sigma)^2}$$

Figure 4.19 shows the total energy, correlated energy, and correlation coefficient E-plane patterns for the 60° biconical antenna driven by a source with a 200Ω resis-

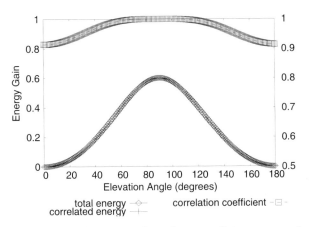

Figure 4.19 Total energy, correlated energy, and correlation coefficient patterns for biconical antenna with baseband Gaussian pulse excitation. Pulse natural width is .428 times the travel time associated with the entire antenna length. From [50].

tance and a baseband Gaussian pulse shape. The dipole is taken as z-directed, and the pulse natural width is $\sigma = 0.428L/c_0$. The template function for the correlated quantities is the waveform radiated in the broadside ($\theta = 90°$) direction, and the reference polarization is vertical.

Because the pulse is relatively long, its energy is concentrated in the lower frequency region in which the biconical dipole radiates primarily a TM_{01} spherical-mode pattern. This is close to the direction-independent-distortion condition; thus, the total and correlated energy patterns are very similar (because almost all of the radiated energy is correlated energy), and the correlation coefficient is fairly close to unity everywhere. This indicates that the shape of the transmitted time-domain pulses is fairly constant with angle, even though there is a null in the energy pattern along the dipole axis.

Figure 4.20 shows the total energy, correlated energy, and correlation coefficient E-plane patterns for the 60° biconical antenna driven by a source with a 200Ω resistance and a cosinusoidal Gaussian burst pulse shape. The pulse natural width is again $0.428L/c_0$. The period of the cosinusoidal function is $0.713L/c_0$. As can be seen, the energy and correlated energy patterns are no longer similar, and the correlation coefficient rolls off sharply as the observation point moves off of the H-plane. This is because much more of the pulse energy lies in the higher frequency range for which the frequency-domain field transfer function is a complicated function of angle. The radiated energy is only well correlated for elevation angles near 90° or near the H-plane of the antenna.

In Figure 4.21, the same quantities are plotted for the same antenna, but with a sinusoidal Gaussian burst source function. Again, the pulse natural width is $0.428L/c_0$, and the period of the sinusoidal function is $0.713L/c_0$. The total energy, correlated

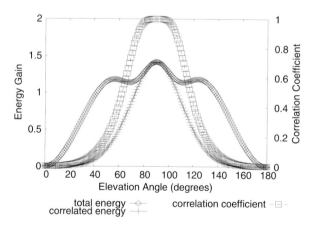

Figure 4.20 Total energy, correlated energy, and correlation coefficient patterns for biconical antenna with cosinusoidal Gaussian burst excitation. Pulse natural width is .428 times the travel time associated with the entire antenna length. Period of cosine is .713 times the same travel time. From [50].

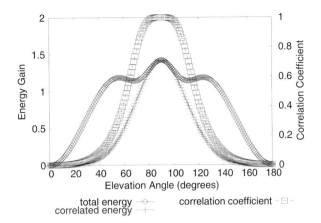

Figure 4.21 Total energy, correlated energy, and correlation coefficient patterns for biconical antenna with sinusoidal Gaussian burst excitation. Pulse natural width is .428 times the travel time associated with the entire antenna length. Period of sine is .713 times the same travel time. From [50].

energy, and correlation coefficient patterns look very similar to those for the cosinusoidal burst.

Figures 4.22 and 4.23 show results for the same antenna and a cosinusoidal and sinusoidal Gaussian burst pulse shape, respectively, but with a lower cosine/sine frequency. The pulse natural width is still $0.428L/c_0$, but the period of the cosine/sine function is $1.07L/c_0$. Here, the correlation is improved over Figures 4.3 and 4.4, but it is still not as good as in Figure 4.2 because the frequency distribution of the spec-

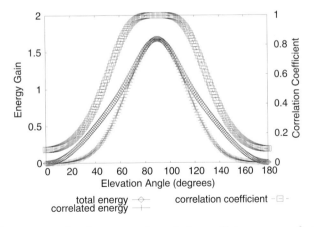

Figure 4.22 Total energy, correlated energy, and correlation coefficient patterns for biconical antenna with cosinusoidal Gaussian burst excitation. Pulse natural width is .428 times the travel time associated with the entire antenna length. Period of cosine is 1.07 times the same travel time. From [50].

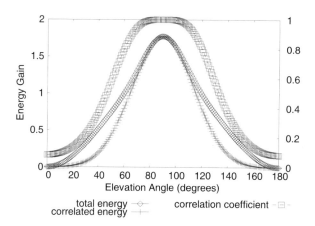

Figure 4.23 Total energy, correlated energy, and correlation coefficient patterns for biconical antenna with sinusoidal Gaussian burst excitation. Pulse natural width is .428 times the travel time associated with the entire antenna length. Period of sine is 1.07 times the same travel time. From [50].

tral content is intermediate between the baseband pulse of Figure 4.2 and the pulses used in Figures 4.3 and 4.4.

In Figures 4.24 and 4.25, the cosinusoidal and sinusoidal Gaussian bursts are again used but with a higher cosine/sine frequency. The pulse natural width remains at $0.428L/c_0$, but the period of the cosine/sine function is $0.429L/c_0$. As expected, the correlation coefficient exhibits considerable variation *m*, and the correlated energy is significantly lower than the total energy. The spectrum in this case lies mainly above the frequency range for which the frequency-domain field transfer function is well behaved.

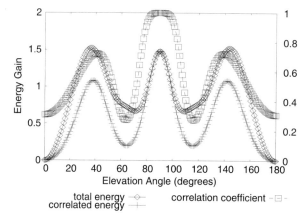

Figure 4.24 Total energy, correlated energy, and correlation coefficient patterns for biconical antenna with sinusoidal Gaussian burst excitation. Pulse natural width is .428 times the travel time associated with the entire antenna length. Period of sine is .429 times the same travel time. From [50].

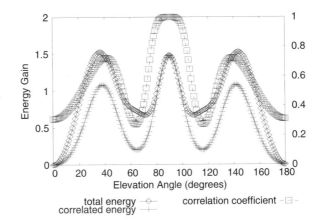

Figure 4.25 Total energy, correlated energy, and correlation coefficient patterns for biconical antenna with cosinusoidal Gaussian burst excitation. Pulse natural width is .428 times the travel time associated with the entire antenna length. Period of cosine is .429 times the same travel time. From [50].

In summary, the biconical antenna is an excellent candidate for many UWB applications, including those in which pulse fidelity is critical. It is probably the most extensively studied UWB antenna design. Its primary drawbacks are its physical size and lack of conformity to flat geometries. Many of the advantages can still be obtained with a bowtie design; however, the pattern is not as stable with frequency. The impact of the pattern difference can be seen clearly in the transfer function magnitude plot (Figure 4.5) shown in Section 4.2.

4.5.6 Shorted Bowtie Antennas

Adding a shorting loop to the outside of a bowtie-type antenna, as shown in Figure 4.26, can improve the performance of the antenna in the region where it is electrically small by stepping up the impedance level. N. Behdad and K. Sarabandi [51] have introduced an antenna in this class, known as the sectorial loop antenna (SLA).

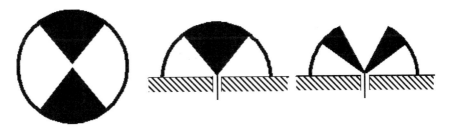

Figure 4.26 Folded bowtie antennas, also called sectorial loop antennas (SLA) [51]. At left, a coupled sectorial loop antenna (CLSA). At center, a half CLSA over a ground plane. At right, a modified CLSA. Typical diameter is 0.37l at the lower edge of the operating band.

These types of antennas can easily cover the UWB band with a return loss greater than 10 dB with considerable size reduction over a conventional bowtie. As one would expect from the lack of azimuthal symmetry, the pattern does change with frequency. For an antenna covering 3.1–10.6 GHz impedance bandwidth, there is up to 10 dB pattern asymmetry in the 3–8 GHz range, and up to 20 dB asymmetry at 10 GHz. Nevertheless, the pattern appears to have more consistent polarization and less pattern variation than similar disc monopoles.

4.5.7 Resistively Loaded Antennas

The beginning of this chapter discussed inherent limitations on bandwidth versus size, remarking that the limits can be evaded by resistively loading the antenna. The resistive loading inhibits "ringing" in the antenna and can lead to improved impedance and pattern bandwidth, and, thus, reduced pulse distortion and time spread. This improvement comes at the cost of significant reduction in energy efficiency; thus, this class of antenna is most suitable for sensing and instrumentation applications and less appropriate for communications applications.

Resistively loaded monopoles and dipoles have been studied extensively, and comprehensive analytical expressions for optimum loading impedance have been given [52–54]. Although the theoretically optimum loading is complex and difficult to realize in practice, very good pulse reproduction has been obtained using purely resistive loading [54].

As a first example, we present numerical simulations of several different loaded linear dipole antennas. The loading is similar to that used in [54], with the form

$$R_t(z) = \frac{R_0}{(L/2) - |z|}$$

where L is the dipole length, z is the position along the wire, and R_t is the resistance per unit length. The source pulse in all cases is Gaussian with an open-circuit source voltage of the form

$$v(t) = e^{-(t/\sigma)^2}$$

As stated in the section on biconical antennas, the relationship between the pulse width parameter σ of the Gaussian pulse and the length L is crucial. In the numerical examples that follow, we have chosen two values for the parameter σ of the Gaussian pulse: $\sigma = 0.15 L/c_0$, and $\sigma = 0.41 L/c_0$.

Figures 4.27 and 4.28 show the time-domain far-zone field for the shorter and longer pulses, respectively, with different values of resistive loading. (The figures could also be viewed as longer and shorter antennas, respectively, for the same pulse length.) The horizontal axis is retarded time, normalized to the transit time for the length of the dipole:

$$\tau = (t - R/c_0)\left(\frac{c_0}{L}\right)$$

where R is the distance of the field point from the antenna and the vertical axis is field-scaled by R to be independent of distance. In all cases, the field is computed in the H-plane of the antenna ($\theta = 90°$).

The electrical size of the antenna is closely related to the optimum loading profile and to the radiated pulse shape. It can be seen that a longer antenna/shorter pulse, as in Figure 4.27, gives a radiated pulse shape that is approximately the first derivative of the source pulse. The short antenna/longer pulse, shown in Figure 4.28, results in a transmitted field that more closely approximates the second derivative of a Gaussian pulse. This is due to the highpass filtering inherent in such a small antenna.

Both Figures 4.27 and 4.28 show that increasing the source impedance (thus, the damping since a small dipole can be approximated as a series resonant circuit) improves the time response, in particular the ringing after the initial pulses. Increasing the resistive loading parameter R_0 also improves the waveform; however, as stated above, this comes at the expense of efficiency, as Table 4.3 shows. None of the cases

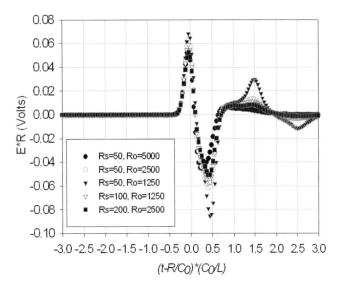

Figure 4.27 Time dependence of transmitted electric field for resistively loaded linear antennas. Incident field is a Gaussian pulse with a width equal to .15 times the travel time associated with the length of the dipole, and an open-circuit peak value of 1 Volt.

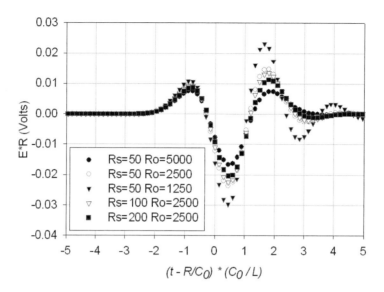

Figure 4.28 Time dependence of transmitted electric field for resistively loaded linear antennas. Incident field is a Gaussian pulse with a width equal to 0.41 times the travel time associated with the length of the dipole. Note that transmitted electric field approximately matches second derivative of Gaussian pulse.

studied had higher than 17 percent energy efficiency; this can be compared to the at least 85 percent efficiency of typical unloaded antennas.

Figures 4.29 through 4.31 show the energy received in a given time window, which is a measure of the shortness of the radiated pulse. The window center is positioned so as to maximize the energy. Generally, higher radiated energy results when the resistive loading R_0 is decreased. However, decreasing the loading increases ringing and, thus, increases the window width needed to capture a given fraction of the radiated energy. This shows that there is a direct trade-off between efficiency and pulse distortion and spreading. Increasing the source resistance R_S improves radiated

Energy Efficiency e_w		
R_S	R_0	Efficiency (%)
50	5,000	3.8
50	2,500	6.3
50	1,250	9.9
100	2,500	10.9
200	2,500	16.7

Table 4.3 Energy Efficiency of a Resistively Loaded Dipole for Different Source Impedances and Loading Profiles

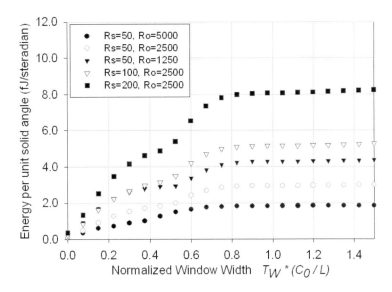

Figure 4.29 Radiated energy per unit solid angle (femtojoules per steradian) versus normalized width of receiving time window, in direction θ = 90°. The source is a Gaussian pulse with a width as in Figure 4.1, but with the amplitude adjusted to have available source energy constant for all cases.

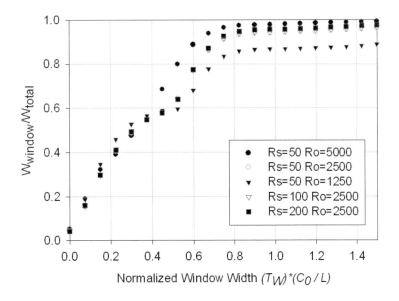

Figure 4.30 Parameter e_T (directional) for resistively loaded dipole in direction θ = 90°, versus normalized window width. It is the fraction of total radiated energy per steradian that is radiated with a given time window width.

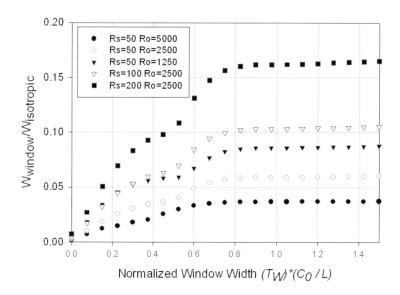

Figure 4.31 Directional time windowed energy efficiency, e_{WT}, versus normalized window width.

energy; however, there are practical limits to how high R_S can go in a microwave transmitter.

The trade-off is clear in Figure 4.30, which shows directional e_T, the fraction of radiated energy that falls within a given time slot (see Section 4.3.3.4). The case with the lowest absolute energy in Figure 4.3 (R_S = 50, R_0 = 5,000) has the greatest fraction of energy captured when the time window is narrow. When R_0 is lowered to 1,250, e_T is reduced to 90 percent of the radiated energy captured, even when the window width is several times the Gaussian pulse width. The remaining 10 percent is in long-duration ringing.

Figure 4.7 shows directional e_{WT}, the total time-windowed energy efficiency. This includes losses due to mismatch and dissipation in the resistive loading. The values of efficiency given in Table 4.3 are the limiting values of e_{WT} as the time window becomes infinitely wide.

The analysis above indicates that somewhat lighter loading than previously proposed provides superior performance. Source damping (high R_S) is shown to be a superior approach to obtaining an improved time response.

R_0 in (4.59) is 5,000Ω, L = 0.2m, and the wire diameter is 0.2 mm. In Figure 4.32, the energy pattern, correlated energy pattern, and correlation coefficient pattern are given for a Gaussian pulse excitation with natural width equal to 0.298 times the transit time associated with the total length of the dipole. The spectrum of this pulse lies in the frequency range over which the dipole exhibits a radiation pattern that is nearly invariant with frequency. Of course, the energy efficiency is very low in this range, as demonstrated by the fact that the energy gain peaks at a value of slightly over 0.02.

UWB Antennas 133

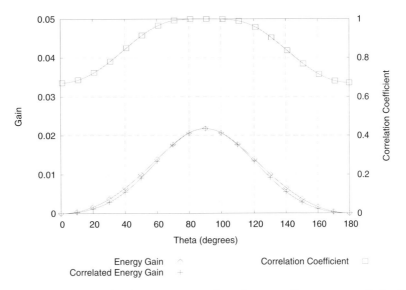

Figure 4.32 Energy gain, correlated energy gain, and correlation coefficient patterns for loaded dipole as specified in [54]. The pulse length is .298 times the transit time associated with the total length of the dipole

The second example (Figure 4.33) is similar to the first, except that the pulse length is halved. As can be seen, more energy is radiated, but pattern is less well correlated.

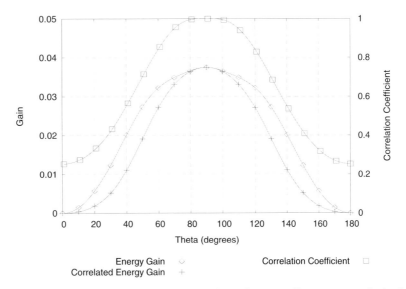

Figure 4.33 Energy gain, correlated energy gain, and correlation coefficient patterns for loaded dipole as specified in [54]. The pulse length is .149 times the transit time associated with the total length of the dipole.

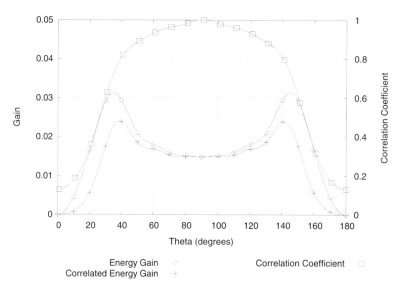

Figure 4.34 Energy gain, correlated energy gain, and correlation coefficient patterns for loaded dipole as specified in [54]. The pulse length is .0745 times the transit time associated with the total length of the dipole.

Finally, the third example (Figure 4.34) involves the same antenna but with a pulse excitation with one-fourth that of the first. Here it can be seen that energy and correlated patterns are no longer similar to those obtained for the long pulse and that the maximum is no longer in the H-plane. Actually, while the correlation of the pattern has degraded significantly, it is still much improved over that of an unloaded dipole [29].

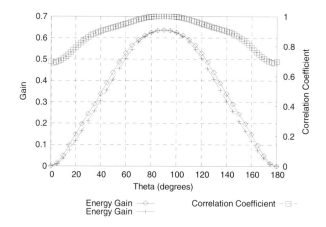

Figure 4.35 Energy gain, correlated energy gain, and correlation coefficient patterns for tapered, loaded dipole made up of 6 wire dipoles loaded as specified in [54]. The pulse length is .0865 times the transit time associated with the total length of the dipole.

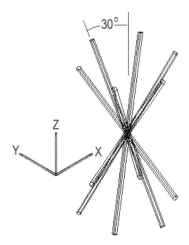

Figure 4.36 Tapered, loaded dipole composed of loaded linear elements.

4.5.8 Tapered, Resistively Loaded Dipole

Because the time-domain performance of the biconical dipole is so much improved over that of the linear dipole, it would be natural to consider a tapered, resistively loaded dipole. In fact, the pulse performance is quite improved over that of the linear loaded dipole.

Here we consider a tapered, loaded dipole comprising six individual loaded wire elements. Each of the six elements has loading identical to that used in the loaded linear dipoles described above.

In Figure 4.35, the energy gain, correlated energy gain, and correlation coefficient patterns are plotted for a loaded tapered dipole antenna, as shown in Figure 4.36. The antenna is actually made up of six linear loaded dipoles in a manner similar to a wire-cage biconical antenna. The opening angle of the cone is 60°. As can be seen, the radiated energy is extremely well correlated, even for this relatively short pulse length. Of course, the energy from longer pulses is even better correlated throughout the pattern. The energy efficiency is 46.8 percent, and the correlated energy efficiency is 44.9 percent.

4.5.9 Frequency-Independent and Quasi-Frequency-Independent Antennas: Log-Periodic Dipole Antennas

Frequency-independent antennas can, in principle, maintain power patterns that are independent of frequency. Practically speaking, such antennas can maintain nearly invariant power patterns over more than a decade of frequency range. They also provide excellent matching efficiency over this broad frequency range. As seen in Equation (4.56), a well-matched antenna with constant power pattern will exhibit a similarly shaped energy pattern. This energy pattern will be independent of pulse shape, provided that the spectra of the pulses are limited to this frequency range.

Quasi-frequency-independent antennas, particularly LPDAs and Archimedian spiral antennas, are also very well suited for UWB applications in many ways. The pattern of a compressed LPDA is particularly appropriate, providing approximately 5–7 dBi gain with perhaps a 20 dB front-to-back ratio. On the other hand, efforts to reduce the size of a frequency-independent antenna further have generally resulted in reduced performance. Thus, frequency-independent antennas are most suitable for applications in which physical size is not an overriding factor.

It is well known that frequency-independent and quasi-frequency-independent antennas such as LPDAs exhibit dispersive behavior [55–60]. Actually, the dispersion of an LPDA is due almost entirely to the reverse-firing feed mechanism [60, 61]. The conventional LPDA feed mechanism, in which every other dipole element is reversed in phase, allows the antenna to be fed from the smaller high-frequency end, while the antenna still radiates primarily in the direction of the high-frequency end. This prevents the detrimental interaction of the larger elements with the smaller elements [61]. Feeding the antenna at the small end then places, at any given frequency, only electrically short elements between the principal active region and the input. If power transfer to the principal active region is effective, only this active region will contribute significantly to the radiation.

Specifically, the feed mechanism causes the phase characteristic of the antenna to be a nonlinear function of frequency. In [57, 58], the phase of the radiated field from a conventional LPDA is shown to vary approximately with frequency as

$$\phi(\omega) = \frac{\pi}{\ln \tau} \ln\left(\frac{\omega}{\omega_1}\right)$$

where τ is the scale factor of the LPDA, and ω_1 is the radian resonance frequency of the shortest dipole element. Thus, the group delay is

$$\tau_g = -\frac{\partial \phi(\omega)}{\partial \omega} = \frac{1}{2 \ln \tau} \frac{1}{f}$$

The distortionless transmission of a time-domain pulse, on the other hand, requires a transfer function that is constant in magnitude and linear in phase variation with frequency. Deviation from linear phase or, equivalently, constant group delay results in distortion of time-domain signals.

It is worth examining the dispersive characteristics of the LPDA more closely. First, it is quite useful to note that the analytical dispersion model proposed by G. J. Burke matches quantitative numerical simulations. The negative slope of the group delay with frequency corresponds to anomalous dispersion that has a strong down or inverse chirping effect on pulses. For example, a Gaussian pulse that is initially unchirped, with all of its Fourier components in phase, will, upon transmission by the LPDA, undergo a down-chirping process. It is clear from the time-domain plot of the response of a typical LPDA that the instantaneous frequency is decreasing sharply with frequency. In principle, it should be possible to use up-chirped pulses in a communications scheme. In any case, it should also be, in principle, possible to design an allpass analog filter with which an LPDA can be equalized.

It is important to differentiate between ringing and chirping. Both processes spread pulses out in time. However, ringing takes place because of the resonance(s) in a narrowband system or device. Such a system almost certainly has poor matching efficiency. Chirping, on the other hand, is not necessarily associated with filtering but, rather, is an allpass characteristic. Thus, chirping results, in principle, in no deterioration of energy transfer. Figure 4.37 gives the time-domain response of an unloaded dipole. As can be seen, the dipole tends to ring at the fundamental resonance as well as at the odd-order harmonics. The ringing in a dipole antenna is somewhat similar to the response of a comb filter.

Numerical simulations of a typical LPDA were performed using the Numerical Electromagnetics Code (NEC), which is an extensively verified, thin-wire, frequency-domain method of moments (MoM) simulator. The frequency-domain analysis was carried out over a sufficient range to allow the time-domain characteristics to be obtained via an inverse Fourier transform. The LPDA design had parameters $\tau = .90$ and $\sigma = .1675$ and employed 23 linear dipole elements. This

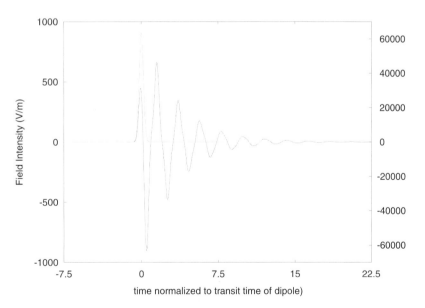

Figure 4.37 Time domain response of unloaded dipole: Gaussian pulse natural width = 3/10 of transit time associated with total dipole length.

combination of scale factor and number of elements gives more than one decade of frequency range between the half-wave resonances of the shortest and longest elements; that is, the structure bandwidth is

$$B_S = \tau^{1-N} = .90^{22} \approx .0985$$

in R. Carrel's nomenclature. This corresponds to a 10:1 frequency range. However, it is important to note that the operating frequency range is significantly less than this range. Normally, for a given frequency, the active region of an LPDA lies almost entirely in front of the half-wave element. In order to avoid truncation effects, the frequency at which the shortest element is one half-wavelength long must be well above the highest operating frequency. That is, the bandwidth of the active region B_{ar} must be significantly larger than 1; it is about 1.6 for the LPDA parameters used here. Thus, one would expect the antenna to provide good quasi-frequency-independent performance over about a 6:1 frequency range. It should be noted that other practical features of a physical LPDA detract from the quasi-frequency-independent operation. In particular, the typical over-/underfeed mech-

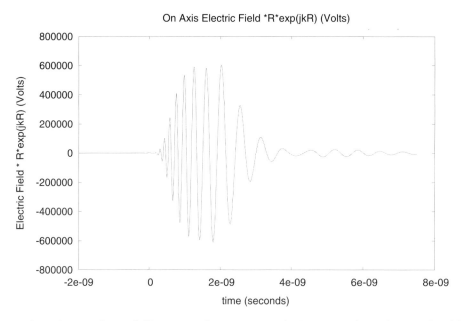

Figure 4.38 On axis electric field response of LPDA to 1-Joule Gaussian pulse with natural width of .0745 nanoseconds. The LPDA has 23 linear dipole elements ranging in length from 7.50 mm to 76.11 mm. The LPDA parameters are $\tau = .90$ and $\sigma = .1675$. These two parameters correspond to the so-called optimum LPDA design criteria given by Carrel. The electric field multiplied by the factor $R \cdot e^{jkR}$ is the actual quantity plotted.

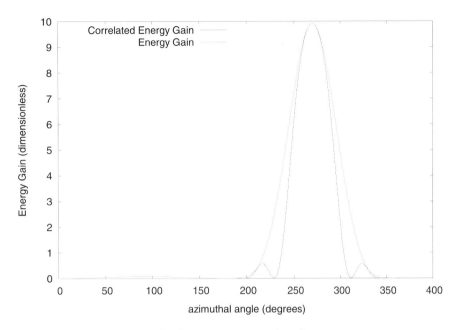

Figure 4.39 Energy gain and correlated energy gain in E-plane for LPDA.

anism used to implement the required interelement phase transposition can limit performance, especially at the upper end of the operating frequency range where it introduces cross-polarized radiation and reduces front-to-back ratio. (Actually, the cross-polarized fields tend to radiate in the direction opposite the bore site.) Also, it is often not possible to maintain scaling of the element diameters over such a broad frequency range. This tends to cause an increase in gain with frequency when the length-to-diameter ratio diminishes with frequency. Here, the length-to-diameter ratio of the elements was maintained at 23.425. Note that the set of LPDA parameters corresponds to a so-called optimum LPDA design in terms of gain for a given length. A resistive load was connected across the feed transmission line at the low-frequency end of the antenna to absorb any power not coupled to the active region(s).

Figure 4.38 shows a typical on-axis electric field response of the LPDA to a Gaussian pulse excitation. As can be seen, the down-chirping effect causes a very long response. However, the correlation of the pulse responses across the principal lobe are quite good, as can be seen in Figures 4.39 and 4.40, which give the energy gain and correlated energy gain. Thus, such an antenna is an excellent candidate for fixed equalization with an allpass filter. In [62], a synthesis procedure for an allpass equalization network and prediction of the equalized LPDA performance are presented.

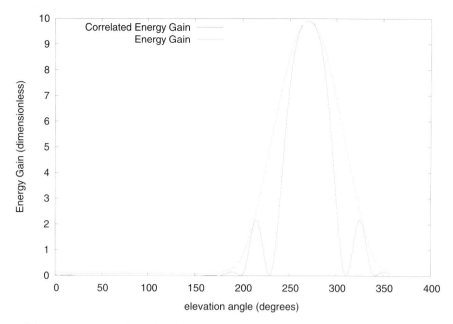

Figure 4.40 Energy gain and correlated energy gain in H-plane for LPDA.

4.6 Conclusion

Commercial UWB applications present some unique challenges to the antenna designer. Their antenna requirements differ from both those of traditional narrowband communications and those of traditional short-pulse applications, such as radar.

From the review in Section 4.1 of the basic restrictions on the bandwidth and size of antennas, it becomes clear that the design of an antenna for a typical commercial UWB platform requires attention to both impedance matching and efficiency limitations on electrically small antennas, as well as to pattern issues associated with moderate-size antennas; that is, the design of a UWB antenna appropriate for a typical mobile platform becomes a highpass impedance-matching problem at the low end of the operating frequency range, and it becomes a challenge to maintain a particular radiation pattern at the high end. The difficulty of simultaneously satisfying both of these requirements cannot be overstated. It is also noted that the large bandwidth makes the maintenance of balance and control for unintentional CM currents on the platform on which the antenna is mounted quite difficult. Finally, it is noted that, occasionally, exaggerated findings of bandwidth and efficiency for an apparent electrical size result from the failure to account for the image of the intended antenna properly. In some cases, the image can be much larger than the "antenna" itself.

In the second section, we introduced antenna transfer functions and showed that signal modification is inherent in a wireless link and not necessarily detrimental. On the other hand, excessive dispersion or pulse distortion that cannot be corrected with fixed equalization places demands on signal processing that necessarily limit system performance.

The performance parameters presented in Section 4.3 will hopefully provide a different viewpoint from which antennas can be evaluated. These parameters are intended to complement traditional antenna parameters as the traditional parameters may not be the most convenient description for UWB applications. We introduced the correlated energy gain and correlation coefficient and the concept of direction-independent distortion. When an antenna exhibits direction-independent distortion, it is possible to correct for it with fixed equalization.

The survey of recent UWB antenna designs emphasized small, low-gain antennas that may be suitable for mobile and portable electronic devices. Undoubtedly, there are numerous antenna designs that we have not covered in detail here. However, it is hoped that our survey has provided some perspective of the current trends in this area. Given the maturity of the research area of electromagnetics, as well as that of antenna design, the reader would be well advised to consider the innumerable designs given in the technical literature and patent databases carefully. It is hoped that the somewhat different perspective afforded by the performance parameters presented in Section 4.3 will provide the reader with a new tool for evaluating these designs for UWB applications.

Finally, a brief and perhaps incomplete discussion of balance and ground-plane independence was presented. This issue is in no way unique to the UWB area; rather, the bandwidth requirements for UWB applications make obtaining ground-plane independence more difficult. Any device, such as a balun placed in the signal path, can affect pulse fidelity. Only a few balun types satisfy the true-time-delay or distortionless transmission requirement. Finally, no balun, even a hypothetical ideal current balun, can balance an antenna that is strongly coupled to a nearby object through its near fields in an asymmetrical fashion.

References

1. Sensor and Simulation Notes, published by the Air Force Weapons Laboratory, Kirtland Air Force Base, New Mexico, available online from Otto-von-Guericke-Universität Magdeburg, Magdeburg, Germany, www-e.uni-magdeburg.de.
2. Stutzman, W. A., and G. L. Thiele. *Antenna Theory and Design*, 2nd ed. New York: Wiley, 1997.
3. Balanis, C. A. *Antenna Theory: Analysis and Design*, 2nd ed. New York: Wiley, 1996.
4. Kraus, J. F., and R. J. Marhefka. *Antennas*, 3rd ed. New York: McGraw-Hill, 2001.
5. Hansen, R. C. "Fundamental Limitations in Antennas." *Proc. IEEE* 69 (February 1981): 170–182.

6. Chu, L. J. "Physical Limitations of Omnidirectional Antennas." *J. Applied Physics* 19 (December 1948): 1163–1176.

7. McLean, J. S. "A Re-examination of the Fundamental Limits on the Radiation Q of Electrically Small Antennas." *IEEE Trans. Antennas and Propagation* 44, no. 5 (May 1996): 672–676.

8. Wheeler, H. A. "Small Antennas." *IEEE Trans. Antennas and Propagation* AP-23 (July 1975): 462–469.

9. Wheeler, H. A. "Fundamental Limitations of Small Antennas." *Proc. IRE* 35 (December 1947): 1479–1484.

10. Harrington, R. F. "Effect of Antenna Size on Gain, Bandwidth, and Efficiency." *J. Res. Nat. Bur. Stand.* 64-D (January/February 1961): 1–12.

11. Collin, R. E., and S. Rothschild. "Evaluation of Antenna Q." *IEEE Trans. Antennas and Propagation* AP-12 (January 1964): 23–37.

12. Fante, R. L. "Quality Factor of General Ideal Antennas." *IEEE Trans. Antennas and Propagation* AP-17 (March 1969): 151–155.

13. Wheeler, Harold A. "The Wide-band Matching Area for a Small Antenna." *IEEE Trans. Antennas and Propagation* 31 (March 1983): 364–367.

14. Fano, R. M. "Theoretical Limitations on the Broadband Matching of Arbitrary Impedances." *J. Franklin Institute* 249 (January 1950): 57–83; (February 1960): 139–155.

15. Harrington, R. F. *Time-Harmonic Electromagnetic Fields.* New York: McGraw-Hill, 1961.

16. Hansen, J. E., ed. *Spherical Near-Field Antenna Measurements.* London: Peter Peregrinus on behalf of the Institution of Electrical Engineers, 1988, 61–69.

17. Sinclair, G. "The Transmission and Reception of Elliptically Polarized Waves." *Proc. IRE* 38 (February 1950): 148–151.

18. Mays, R. P. "A Summary of the Transmitting and Receiving Properties of Antennas." *IEEE Antennas and Prop. Magazine* 42, no. 3 (June 2000): 49–53.

19. Lamensdorf, D., and L. Susman. "Baseband-Pulse-Antenna Techniques." *IEEE Antennas and Prop. Magazine* 36, no. 1 (February 1994): 20–30.

20. Qing, X., and Z. N. Chen. "Transfer Functions Measurement for UWB Antenna." *IEEE Antennas and Propagation Symposium* 3 (June 2004): 2532–2535.

21. Pozar, D. M. "Closed-Form Approximations for Link Loss in a UWB Radio System Using Small Antennas." *IEEE Trans. Antennas and Propagation* 51, no. 9 (September 2003): 2346–2352.

22. Ishigami, S., H. Takashi, and T. Iwasaki. "Measurements of Complex Antenna Factor by the Near-Field 3-Antenna Method." *IEEE Trans. Electromagnetic Compatibility* 38, no. 3 (August 1996): 424–432.

23. Shlivinski, A., E. Heyman, and R. Kastner. "Antenna Characterization in the Time Domain." *IEEE Trans. Antennas and Propagation* 45, no. 7 (July 1997): 1140–1149; *see also* correction in *IEEE Trans. Antennas and Propagation* 45, no. 8 (August 1997): 1323.

24. Farr, E. G., and C. E. Baum. "Extending the Definitions of Antenna Gain and Radiation Pattern into the Time Domain." in *Sensor and Simulation Notes*, Note 350, Air Force Research Laboratory, 1992.

25. B. Scheers, M. Acheroy, and A. Vander Vorst. "Time Domain Simulation and Charactisation of TEM Horns using a Normalized Impulse Response." *IEEE Proc. Microwaves, Antennas and Propagation* 147, no. 6 (December 2000): 463–468.

26. Sörgel, W., F. Pivit, and W. Wiesbeck. "Comparison of Frequency Domain and Time Domain Measurement Procedures for Ultra Wideband Antennas." *Proc. 25th Annual Meeting and Symposium of the Antenna Measurement Techniques Association* (AMTA '03), Irvine, California (October 2003): 72–76.

27. Pozar, D. M. "Waveform Optimizations for Ultra-Wideband Radio Systems." *IEEE Trans. Antennas and Propagation* 51, no. 9 (September 2003): 2335–2345.

28. Pozar, D. M. "The Optimum Feed Voltage for a Dipole Antenna for Pulse Radiation." *IEEE Trans. Antennas and Propagation* (July 1983): 563–569.

29. McLean, J. S., H. Foltz, and R. Sutton. "Pattern Descriptors for UWB Antennas." *IEEE Trans. Antennas and Propagation* 53, no. 1 (January 2005): 553–559.

30. McLean, J. S., H. D. Foltz, and R. Sutton. "The Effect of Frequency-Dependent Radiation Pattern on UWB Antenna Performance." *2004 IEEE APS Symposium Digest*, Monterey, California, June 21, 2004.

31. Scott, W. G., and K. M. Soohoo. "A Theorem on the Polarization of Null-Free Antennas." *IEEE Trans. Antennas and Propagation* AP-14 (September 1966): 587–590.

32. Kerkhoff, A., and H. Ling. "A Parametric Study of Band-Notched UWB Planar Monopole Antennas." *2004 IEEE Antennas and Propagation Society International Symposium Digest* 2 (June 2004): 1768–1771.

33. Johnson, W. A. "The Notch Aerial and Some Applications to Aircraft and Installations." *Proc. IEE* 102, part B (March 1955): 211–218.

34. Sawaya, K., T. Ishizone, and Y. Mushiake. "A Simplified Expression for the Dyadic Green's Function for a Conducting Half Sheet." *IEEE Trans. Antennas and Propagation* 29, no. 5 (September 1981): 749–756.

35. Venkatesan, J. B. "Investigation of the Double-Y Balun for Feeding Pulsed Antennas." Ph.D. dissertation, School of Electrical and Computer Engineering, Georgia Institute of Technology, July 2004.

36. Oltman, G. "The Compensated Balun." *IEEE Trans. Microwave Theory and Techniques* MTT-14 (March 1966): 112–119.

37. Phelan, H. Richard. "A Wide-band Parallel-Connected Balun." *IEEE Trans. Microwave Theory and Techniques* MTT-18, no. 8 (March 1966): 259–263.

38. McLean, J. S. "Balancing Networks for Symmetric Antennas—1: Classification and Fundamental Operation." *IEEE Trans. Electromagnetic Compatibility* 44, no. 4 (November 2002): 503–514.

39. McLean, J. S. "Balancing Networks for Symmetric Antennas—2: Practical Implementation and Modeling." *IEEE Trans. Electromagnetic Compatibility* 44, no. 4 (November 2002): 503–514.

40. Wong, K.-L., and S.-L. Chien. "Wide-band Cylindrical Monopole Antenna for Mobile Phone." *IEEE Trans. Antennas and Propagation* 53, no. 8 (August 2005): 2756–2758.

41. Wong, K.-L., and S.-L. Chien. "Wide-band Omnidirectional Square Cylindrical Metal-Plate Monopole Antenna." *IEEE Trans. Antennas and Propagation* 53, no. 8 (August 2005): 2758–2761.

42. Kerkhoff, A., and H. Ling. "Design of a Planar Monopole Antenna for Use with Ultra-Wideband (UWB) Having a Band-Notched Characteristic." *2003 IEEE Antennas and Propagation Society International Symposium Digest* 1 (June, 2003): 830–833.

43. Yoon, H., H. Kim, K. Chang, Y. J. Yoon, and Y.-H. Kim. "A Study on the UWB Antenna with Band-Rejection Characteristic." *2004 IEEE Antennas and Propagation Society International Symposium Digest* 2 (June 2004): 1784–1787.

44. Goubau, G. "Multi-Element Monopole Antennas." *Proc. ECOM-ARO Workshop on Electrically Small Antennas*, Fort Monmouth, New Jersey (May 1976): 63–67.

45. Cobos, L., H. Foltz, and J. McLean. "A Modified-Goubau-Type Antenna with Two Octaves of Impedance Bandwidth." *2004 IEEE Antennas and Propagation Symposium* 3 (June 2004): 3051–3054.

46. Friedman, C. H. "Wide-band Matching of a Small Disk-Loaded Monopole." *IEEE Trans. Antennas and Propagation* AP-33 (October 1985): 1142–1148.

47. Ravipati, C. B., and C. J. Reddy. "Low Profile Disk and Sleeve Loaded Monopoles." *2005 IEEE Antennas and Propagation International Symposium Digest*, Washington, DC, July 2005.

48. Bevensee, R. M. *A Handbook of Conical Antennas and Scatterers*. New York: Gordon and Breach Science Publishers, 1973.

49. Brown, G. H., and O. M. Woodward. "Experimentally Determined Radiation Characteristics of Conical and Triangular Antennas." *RCA Review* 13 (December 1952): 425–452.

50. McLean, J. S., H. D. Foltz, and R. Sutton. "The Effect of Frequency-Dependent Radiation Pattern on UWB Antenna Performance." *2004 IEEE Antennas and*

Propagation International Symposium Digest, Monterrey, California, June 21, 2004.

51. Behdad, N., and K. Sarabandi. "A Compact Antenna for Ultra Wide-band Applications." *IEEE Trans. Antennas and Propagation* 53, no. 7 (July 2005): 2185–2192.

52. Kanda, M. "Transients in a Resistively Loaded Linear Antenna Compared with Those in a Conical Antenna and a TEM Horn." *IEEE Trans. Antennas and Propagation* AP-28, no. 1 (January 1980): 132–136.

53. Wu, T. T., and R. W. P. King. "The Cylindrical Antenna with Non-Reflecting Resistive Loading." *IEEE Trans. Antennas and Propagation* 13, no. 3 (November 1965): 998.

54. Kanda, M. "A Relatively Short Cylindrical Broadband Antenna with Tapered Resistive Loading for Picosecond Pulse Measurements." *IEEE Trans. Antennas and Propagation* AP-26, no. 3 (May 1978): 439–447.

55. Knop, C. M. "On Transient Radiation from a Log-Periodic Dipole Array." *IEEE Trans. Antennas and Propagation* AP-18, no. 6 (November 1970): 807–809.

56. Yatskevich, V. A., and L. L. Fedosenko. "Antennas for Radiation of Very Wide-band Signals." *Radio Electronics and Communication Systems* 29, no. 2 (1986): 62–66.

57. Burke, G. J. "Evaluation of Modified Log-Periodic Antennas for Transmission of Wide-band Pulses." *Proc. 7th Annual Review of Progress in Applied Computational Electromagnetics*, Monterey, California, January 1991.

58. Burke, G. J. "Evaluation of Modified Log-Periodic Antennas for Pulse Transmission." *Lawrence Livermore National Laboratory Report UCRL-ID-106868,* April 1991.

59. Excell, P. S., A. D. Tinniswood, and R. W. Clarke. "An Independently Fed Log-Periodic Antenna for Pulsed Radiation." *IEEE Trans. Electromagnetic Compatibility* 41, no. 4 (November 1999): 344–349.

60. McLean, J. S., and R. Sutton. "A Log-Periodic Dipole Antenna Employing a Forward-Firing Feed Mechanism for Pulse Radiation." *IEEE Intl. Symposium on Antennas and Prop.*, Columbus, Ohio, 2003.

61. Isbell, D. E. "Log Periodic Dipole Arrays." *IRE Trans. Antennas and Propagation* AP-8 (May 1960): 260–267.

62. Hirschmüller, E., and G. Mönich. "Realization of All-Pass-Networks for Linearizing Logarithmic-Periodic Dipole Antennas." *EuroEM 2004,* Magdeburg, Germany, July 2004.

5

Direct-Sequence UWB
by Michael McLaughlin

5.1 Direct-Sequence UWB

Ever since the advent of consumer communications with the voiceband modem, engineers have been striving to pack higher and higher bit rates into every piece of spectrum they are allocated. It started with the telex at 5 to 50 bps over a 3 kHz band, went through 300 bps with Bell 103 and V.21, 1,200 bps with Bell 212 and V.22, finally culminating in V.34 at 33.6 kbps, which, to this day, is one of the most spectrally efficient communications schemes ever devised. The focus moved on to wireless, and the same race with the same goal of greater and greater spectral efficiency continued. There was lot more spectrum available, but it was all soon eaten up and filled to capacity.

Then, along came UWB with a fresh approach. When it arrived on the scene, the philosophy of UWB was completely different. Now, there is no need to be spectrally efficient. If you send very narrow pulses, the spectrum is so wide that you can have as much of it as you like. These narrow pulses have very low power in any particular band, so interference is minimized. But even with this huge bandwidth, the hunger for high bit rates can never be satisfied, and no matter how wide the spectrum is, if you do not send enough pulses per second, you cannot get enough bits per second. This need for higher bit rates led to the development of the Direct-Sequence UWB (DS-UWB) approach. Eventually, if you send pulses at a high enough pulse rate, you end up sending them as a train in every available chipping slot, and that is the principle of a DS-UWB scheme.

Direct sequence (i.e., sending a train of ones and zeros to represent one or more bits) was used in earlier communications systems. Usually, the bits were modulated onto a carrier where, for example, a one is represented by one phase shift, and a zero is represented by the opposite phase shift. With UWB, it was natural to represent a one with a positive pulse and a zero with a negative pulse, but it was also natural to have a gap between pulses. So, these three natural possibilities meant that ternary sequences were not only possible but actually very easy to generate. A ternary sequence, where each chip has three possibilities, greatly increases the number of possible sequences.

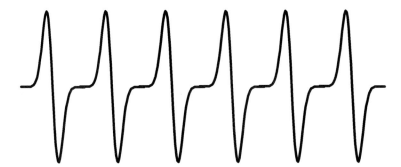

Figure 5.1 Pulse train which could, for example, represent binary zero.

DS-UWB is a modulation scheme where the symbols representing the information to be transmitted consist of binary or ternary sequences of pulses. For example, a sequence of six pulses, all with the same polarity, could represent an input binary one, and six pulses of alternating polarity could represent a binary zero. The receiver would then have the task of deciding which of these two possibilities was transmitted in order to determine whether a one or a zero was input to the modulator.

In a DS-UWB system, the sequences of pulses in the example above would be represented by either binary sequences or ternary sequences. The train of positive-going pulses would be represented by the binary sequence +1 +1 +1 +1 +1 +1, and the train of alternating pulses by the binary sequence +1 −1 +1 −1 +1 −1. In some literature, a zero is used to represent the negative pulse, and a one is used to represent the positive pulse; so, in this example, that would be 11111 and 101010. One disadvantage of this notation is that ternary sequences cannot be represented. Another is that it implies that there is a dc offset, which is not present in practice.

In other literature, the positive pulse is represented by a "+" sign and the negative pulse by a "−" sign. This has the advantage that a ternary code can be represented by introducing a zero into the alphabet. This latter notation will be used in this chapter. The two sequences in the example above then are ++++++ and +−+−+−. A ternary code (e.g., +0+0−−) can also be represented. This particular ternary sequence would represent the pulse sequence in Figure 5.1.

5.2 Binary Signaling with DS-UWB

It would be perfectly possible to use the example sequences given above to represent binary ones and zeros, but in practice, when binary signaling, or binary phase-shift keying (BPSK), is used in DS-UWB, antipodal signals are used for the one and the zero (i.e., the sequence used for binary one is the negative of the sequence used for binary zero). This is because antipodal signals are the optimal signals to use for binary signaling in terms of giving the lowest bit error rate for a given signal-to-noise ratio (SNR) (see, e.g., [1], Section 4.2.1).

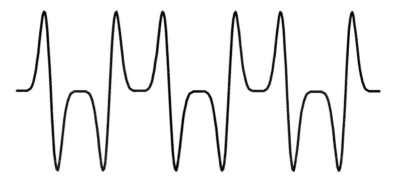

Figure 5.2 Pulse train which could, for example, represent binary one.

Also, as we shall see during the course of this chapter, the particular example spreading codes used above are not well suited for binary signaling in an UWB channel.

5.3 *M*-ary Biorthogonal Signaling with DS-UWB

An extension of binary signal is *M*-ary orthogonal signaling (see, e.g., [1], Section 4.2.2). With this type of modulation, instead of just two different pulse sequences representing one information bit, there are M sequence signals representing $\log_2 M$ information bits. All of these signals are orthogonal to each other. Two sequences are said to be orthogonal if their inner product is zero (i.e., if the sum of the product of their corresponding elements is zero). The sequence in Figure 5.1 is orthogonal to the sequence in Figure 5.2.

One variation of this, which gives improved performance, is where simplex sequences are used instead of orthogonal sequences. To construct a set of M simplex sequences, the mean of all the orthogonal sequences is subtracted from each sequence. This has the effect of translating the origin of the signals to zero, which reduces the transmit power without changing the distance between the sequences.

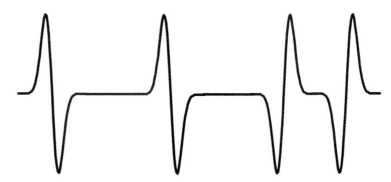

Figure 5.3 Example ternary sequence represented by +0+0--.

Another variation of *M*-ary orthogonal signaling is *M*-ary biorthogonal signaling (MBOK) (see, e.g., [1], Section 4.2.3). In MBOK, the signal alphabet contains *M*/2 orthogonal sequences, plus another *M*/2 sequences that are simply the negatives of these sequences.

This is also better than *M*-ary orthogonal signaling because it allows one more bit to be transmitted per symbol, without increasing either the transmit power or the minimum distance between sequences. Of the three variations discussed here, MBOK is the most commonly used for direct-sequence modulation.

5.3.1 Coding Gain of MBOK

MBOK has an inherent coding gain over uncoded BPSK. To see why this is so, consider a 16-ary MBOK system (16-BOK).

A 16-element biorthogonal alphabet can be constructed by using a length-8 Walsh-Hadamard matrix and its negative as shown in Table 5.1.

This type of MBOK allows four bits to be transmitted for every eight chips. All the sequences have the same squared Euclidean distance (SED) from every other sequence, except for the one sequence that is their own negative, where the SED is double. The minimum squared Euclidean distance (MSED) for these sequences is 16.

To construct an equivalent alphabet for a BPSK system where eight chips encode four bits, every two chips encode one bit and are either +1 +1 or –1 –1, giving the results shown in Table 5.2.

1	1	1	1	1	1	1	1
1	–1	1	–1	1	–1	1	–1
1	1	–1	–1	1	1	–1	–1
1	–1	–1	1	1	–1	–1	1
1	1	1	1	–1	–1	–1	–1
1	–1	1	–1	–1	1	–1	1
1	1	–1	–1	–1	–1	1	1
1	–1	–1	1	–1	1	1	–1
–1	–1	–1	–1	–1	–1	–1	–1
–1	1	–1	1	–1	1	–1	1
–1	–1	1	1	–1	–1	1	1
–1	1	1	–1	–1	1	1	–1
–1	–1	–1	–1	1	1	1	1
–1	1	–1	1	1	–1	1	–1
–1	–1	1	1	1	1	–1	–1
–1	1	1	–1	1	–1	–1	1

Table 5.1 Example 16-ary Biorthogonal Sequence Alphabet Allowing Four Bits per Sequence

−1	−1	−1	−1	−1	−1	−1	−1
−1	−1	−1	−1	−1	−1	1	1
−1	−1	−1	−1	1	1	−1	−1
−1	−1	−1	−1	1	1	1	1
−1	−1	1	1	−1	−1	−1	−1
−1	−1	1	1	−1	−1	1	1
−1	−1	1	1	1	1	−1	−1
−1	−1	1	1	1	1	1	1
1	1	−1	−1	−1	−1	−1	−1
1	1	−1	−1	−1	−1	1	1
1	1	−1	−1	1	1	−1	−1
1	1	−1	−1	1	1	1	1
1	1	1	1	−1	−1	−1	−1
1	1	1	1	−1	−1	1	1
1	1	1	1	1	1	−1	−1
1	1	1	1	1	1	1	1

Table 5.2 *Equivalent Length-8 Sequences for a BPSK System Allowing Four Bits per Sequence*

By examining these sequences, it is easy to see that, unlike the 16-BOK case, the distance to other sequences varies a lot. It is either 8, 16, 24, or 32. The MSED for these sequences is therefore eight. This is 3.01 dB worse than for 16-BOK. This sets an upper bound on the coding gain of 16-BOK modulation at 3.01 dB. The simulation results shown in Figure 5.4 show that, in practice, at an error rate of 10^{-5}, the coding gain for 16-BOK is about 2.5 dB, and for 8-BOK, it is a little over 1.5 dB.

5.3.2 Combining MBOK with Convolutional Coding

By applying convolutional coding, a coding gain can be achieved over uncoded BPSK because, by adding redundant bits, it is possible to disallow certain combinations of bits, the combinations that have low SED, from the valid combinations. For example, in the uncoded BPSK example above, the MSED is eight. If a good convolutional code were used, it would disallow combinations of bits that differed by only one bit error (e.g., it would only allow sequences that differed by, say, 32 or 24). In the BPSK case, lots of combinations have an SED of either 32 or 24, but if a convolutional code is used with MBOK, for each MBOK sequence there is only one other sequence where the SED is as high as 32. The SED to the other 15 sequences is only 16. For this reason, the coding gain that results when MBOK is combined with convolutional coding is not additive. For example, if a convolutional code with a gain of 4.0 dB were used with 16-BOK, the overall coding gain would be quite a bit less than 7.01 dB. It is quite simple to draw a trellis diagram for a convolutional code used with MBOK rather than BPSK and to replace the partial path distances with

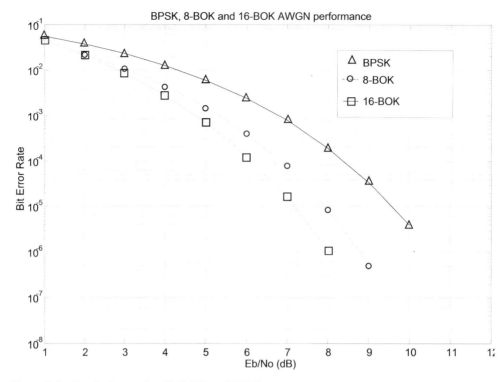

Figure 5.4   Simulation results of MBOK and BPSK.

the correct ones for MBOK. This usually results in an overall MSED close to, but actually smaller than, the MSED that the convolutional code alone gives.

5.4 Properties of Good Codes

A transmit-receive block diagram for a matched DS-UWB receiver can be simplified to blocks shown in Figure 5.5.

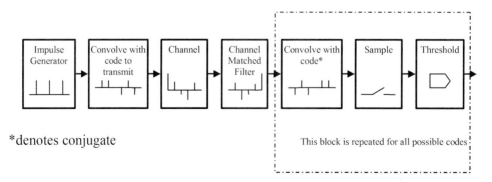

Figure 5.5   Simplified transmit-receive block diagram for a matched DS-UWB receiver.

Direct-Sequence UWB

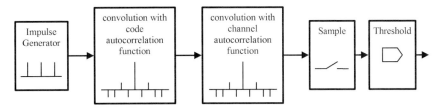

Figure 5.6 Equivalent of transmitter and matched receiver block diagram.

The order of the linear operations can be changed. The convolution with the code and, subsequently, with its conjugate can be replaced by convolution with the autocorrelation function of the code. The same can be done with the channel and the channel-matched filter to get the equivalent block diagram shown in Figure 5.6.

It can be seen that the nice clean impulses of the transmitter are spread out, both by the side lobes of the code autocorrelation function and by the side lobes of the channel autocorrelation function.

5.4.1 Intersymbol Interference

This spreading out of the transmitter impulses causes each of the symbols to spread into the symbols on either side. This is known as intersymbol interference (ISI). It will be smaller for channels with low delay spread (i.e., channels where the side lobes are small and stay close to the central lobe). ISI can also be minimized by a good choice of code (i.e., a code whose autocorrelation function has a large central peak and a small amount of energy in the side lobes).

5.4.2 Peak-to-Average Ratio

Two types of peak-to-average ratio (PAR) are important to keep under control in any UWB modulation scheme, namely, temporal PAR and spectral PAR.

5.4.2.1 Temporal PAR

The temporal PAR is the peak-to-average ratio in the time domain. In any communications system, there are nearly always limits on the average power that may be transmitted. These are often imposed by government regulatory bodies such as the FCC in the United States. In addition to this limit, there are physical limits to the amount of peak voltage or current that can be generated by the hardware for various reasons. The limiting factors include the following:

- The analog voltage slew rate
- The maximum power-supply voltage
- Nonlinearity problems increase with higher voltages
- The process technology used

Consequently, it is desirable to minimize the instantaneous peak output power for a given mean output power (i.e., to keep the temporal PAR as low as possible).

5.4.2.2 Spectral PAR

UWB spectral limits are set in terms of spectral power density. The most useful part of the allowed transmit spectrum is flat. For a signal to comply with these limits, its worst-case spectral density must stay below this straight line. If the signal spectrum has a peak anywhere, this peak must be kept below a certain limit by attenuating it as well as the rest of the signal. This effectively means that the whole signal must be penalized by any amount that the spectral PAR exceeds 0 dB. For this reason, the spectral PAR also needs to be kept as low as possible.

The spectral PAR of a DS-UWB signal is determined by the following:

1. The spectral PAR of the codes. This is especially important in a binary signaling mode. For MBOK with a large M, codes are being randomly chosen, depending on the input data, from a large selection of codes, all with different spectra. This means that the average spectrum is flatter than the spectrum of the MBOK individual codes.

2. The spectral PAR of the transmit pulse. This pulse shape can be chosen to offset other factors in order to keep as close to the regulatory requirements as possible.

3. The spectral properties of the analog hardware. This includes the filters and the antenna.

4. The frequency resolution and averaging time of the measuring equipment. This can have a big influence on the measured spectrum; for example, relatively short measurement times will tend to whiten the spectrum by adding together the spectra of a lot of uncorrelated segments of longer repeating sequences.

The spectral PAR can be lowered by scrambling the pulse polarities by multiplying them by a much longer pseudorandom sequence. This has the effect of whitening the spectrum. This procedure has other side effects, which may or may not be important. For example, it makes the channel look different after the receiver has undone this scrambling.

5.4.3 Autocorrelation Properties

We have seen how ISI is worse for codes where the power of the autocorrelation function is spread out to the side lobes rather than being concentrated at the central peak. There are two types of autocorrelations, namely, aperiodic autocorrelation and periodic autocorrelation.

5.4.3.1 Aperiodic Autocorrelation Function

The aperiodic autocorrelation function is found by taking the code and surrounding it by an infinite number of zeros and then finding the cross correlation of this new construct with itself. This is often simply referred to as the autocorrelation function.

5.4.3.2 Periodic Autocorrelation Function

The periodic autocorrelation function (PACF) of a code is designed to measure the autocorrelation properties of a sequence that consists of a single code repeated periodically. It is found by correlating the code sequence with its own periodic extension. It is found by calculating the cross-correlation of a single instance of a code with the same code repeated an infinite number of times. The result is a periodic sequence with a period equal to the length of the code.

5.4.3.3 Golay Merit Factor

The Golay merit factor (GMF) measures the quality of the aperiodic autocorrelation function of a sequence by calculating the ratio of the square of the central peak in the autocorrelation function to the sum of the squares of the off-peak side lobes. This measure was introduced by M. J. E. Golay in [2].

The greater the central peak of the autocorrelation, the better the noise immunity because this is what is sampled in the receiver. The side lobes of the autocorrelation function do not extend all the way to the center of the next code. They stop just short of this, so they can never actually cause ISI directly, but codes with low power in the side lobes of their autocorrelation function still have greater immunity to ISI. This is because the signal in these side lobes is spread out further by the channel, and it does not have to be spread as far as the main lobe to interfere with neighboring symbols.

The binary sequence with the highest known GMF is the length-13 Barker sequence. Golay himself always regarded this sequence, whose merit factor is 14.08, as a "singularity of nature whose goodness would never again be attained." Codes with high GMF will usually also have flat spectral properties.

5.4.4 Cross-Correlation Properties

The cross-correlation of codes is another very important property of direct-sequence codes. Take the example of an MBOK receiver looking for code B with a correlator matched to code A. In this case, in Figure 5.6, the code autocorrelation function can be replaced by the cross-correlation function of codes A and B. Ideally, at the correct sampling point (i.e., at the center of the cross-correlation function), the cross-correlation will be zero. This minimizes the chance of incorrectly decoding code A as code B (or as the negative of code B). Two or more codes with this property are said to be orthogonal. Ideally, all the code words in an MBOK system are mutually orthogonal, but systems are sometimes designed where the codes are only approximately orthogonal.

5.4.4.1 Periodic Cross-Correlation

The periodic cross-correlation of two codes is the cross-correlation of one code with the periodic extension of the other. This is not as important in direct-sequence communications systems as the aperiodic cross-correlation since the codes may not be sent repetitively.

5.4.4.2 Aperiodic Cross-Correlation

The aperiodic cross-correlation of two codes is the cross-correlation of a single instance of one code with a single instance of another. This property is especially important in code division multiple access (CDMA) systems, where different codes are used for different channels. In this case, the lower the cross-correlation, the less likely it is that one code will be mistaken for another. It is more important in direct-sequence communications systems than periodic cross-correlation since it treats the codes as isolated events, but it does not tell the whole story either, because, as in the example above, this cross-correlation function is modified by the channel and by the channel matched filter, which tends to give the overall response a Gaussian distribution.

5.5 Binary Codes

This section introduces some well-known binary codes used in direct-sequence systems.

5.5.1 Walsh Codes

The Walsh codes, or as they are also sometimes known, the Walsh-Hadamard codes, comprise one the best-known families of codes. The code words of a Walsh code are all the Walsh functions of a given sequence length, say n. When these Walsh functions are arranged as the rows of a matrix, they form an $n \times n$ Hadamard matrix, H_n. A Hadamard matrix has the property $H_n \cdot H_n^T = nI_n$, where I_n is the $n \times n$ identity matrix, and H_n^T is the transpose of H_n.

A $2n \times 2n$ Walsh-Hadamard matrix, W_{2n}, may be formed recursively from a $n \times n$ Walsh-Hadamard matrix, W_n, as follows:

$$W_{2n} = \begin{bmatrix} +W_n & +W_n \\ +W_n & -W_n \end{bmatrix}$$

e.g., if

$$W_n = \begin{bmatrix} + & + \\ + & - \end{bmatrix} \Rightarrow \quad W_{2n} = \begin{bmatrix} + & + & + & + \\ + & - & + & - \\ + & + & - & - \\ + & - & - & + \end{bmatrix}$$

The code words of a Walsh code, as with all Hadamard codes, are mutually orthogonal, which allows them to be used for MBOK signaling. A drawback of Walsh codes is that the code words have low GMF (i.e., poor autocorrelation properties). There are, however, ways to transform Walsh codes into codes that have better GMF and retain the property of being mutually orthogonal.

5.5.2 Barker Sequences

A Barker sequence is a binary sequence in which the off-peak terms of its autocorrelation function are either 1, 0, or −1. Table 5.3 gives all of the known Barker sequences (excluding negations and reversals); it is thought that no others exist.

The length-11 and length-13 codes have GMFs of 12.1 and 14.08, respectively. These are the two highest GMFs of any binary codes.

5.5.3 *m*-Sequences

Maximum-length shift-register sequences, or *m*-sequences, are generated by multiplying together the values in certain stages of a shift register containing positive ones and negative ones and feeding the output back into the shift register. The result is an *m*-sequence if its PACF is n at one sample period and −1 everywhere else, where the length of the sequence is $n = 2^m - 1$, and m is the number of elements in the shift register. This PACF is close to an impulse, which means that repeated *m*-sequences provide a good channel estimation signal. Provided that the impulse response of the channel is shorter than the sequence, a good estimate of the channel impulse response can be found by convolving the receive signal with the conjugate of the *m*-sequence (i.e., by correlating with the *m*-sequence).

5.5.4 Kasami Sequences

Kasami sequences are a set of $2^{m/2}$ sequences of length $2^m - 1$, where m is an even integer. They can be generated from *m*-sequences. Kasami sequences have the prop-

Length	Code	GMF (see Section 5.7)
3	++−	4.5
4	+−++, +−−−	4.0, 4.0
5	+++−+	6.25
7	+++−−+−	8.167
11	+++−−−+−−+−	12.1
13	+++++−−++−+−+	14.083

Table 5.3 *All Known Barker Sequences*

erty that their worst-case periodic cross-correlation is optimal for a set of that number and length (see [1], p. 836).

5.5.5 Gold Sequences

Gold sequences are also usually generated from m-sequences. There are $2^m + 1$ sequences in a set, each of length $2^m - 1$ (i.e., there are approximately twice as many Gold sequences of a given length as Kasami sequences). They have quite good periodic cross-correlation in that the worst-case cross-correlation is approximately 2 times optimal for even m and 1.41 times optimal for odd m (see [1], pp. 834–835). Gold sequences are also useful because an orthogonal set of 2^m sequences, known as orthogonal Gold codes, can be constructed by just appending a –1 to each of the sequences.

5.6 Ternary Codes

The binary codes we have been considering so far are codes whose chips may take on the value +1 or –1. In a DS-UWB system, these values correspond to either a positive or a negative pulse. DS-UWB signaling is ideally suited to using ternary codes where the chips may take on a zero value (i.e., the pulse may be absent). In modulation systems suitable for ternary codes, they have advantages over binary codes.

5.6.1 Increased Number of Available Codes

The extra option for each pulse means that there are very many more ternary codes than binary codes at any given length. For example, there are 2^4, or 16, different length-4 binary codes as opposed to 3^4, or 81, different length-4 ternary codes. The difference in quantity gets even more marked the longer the codes are; for example, there are 2^{16}, or 65,536, different length-16 binary codes, whereas there are 3^{16}, or 43,046,721, different length-16 ternary codes. This large increase in the number of available codes makes it much more likely that a code with desirable properties exists for a given code length. The flipside of the same coin is that automated searches for good codes have a much larger search area to cover.

5.6.2 Autocorrelation Properties of Ternary Codes

To illustrate how the increased number of codes leads to improved code properties, take the example of Barker sequences. If you include reversals but not negations, there are only eight binary Barker sequences with a GMF of six or better, two each at lengths 5, 7, 11, and 13. Ternary Barker sequences (i.e., ternary sequences whose maximum off-peak autocorrelation magnitude is 1.0) are much more abundant. An exhaustive search of the 43,046,721 ternary codes of length 16 reveals that, excluding negations, there are in fact 738 ternary Barker codes with a GMF of six or better. Four of these have a GMF of 14.08: the two binary Barker sequence, this length-15 code, ++++0--++-0+-+-, and its reversal. Also, unlike with binary codes, as the length increases, the properties get better. There are many ternary codes with a GMF of greater than 14.08. The highest merit factor found so far for a ternary code is 20.06 for a length-23 code [3].

5.6.3 Impulse Radio and Ternary Codes

One of the first forms of UWB was called impulse radio. In this variation, positive and negative impulses are transmitted, separated by variable and relatively long pauses. In a technique known as Time Hopping, the lengths of the pauses are often decided by a pseudo random sequence. DS-UWB and impulse radio have a lot in common; for example, using a single positive or negative impulse to represent an information bit is the same as using a ternary sequence where one chip is a single one or minus one, and all the other chips are zero. Time hopping can also be represented by changing the position of the one or minus one in the code.

An impulse like this has ideal autocorrelation properties and an infinite GMF, and so its immunity to ISI could not be better. Similarly, its spectral PAR is 1.0 (i.e., ideal).

A disadvantage of this type of code is the temporal PAR, which is very large for a narrow pulse with a low repetition frequency. This could be a problem for systems with longer periods between pulses (i.e., long codes). The off-peak cross-correlation of two interfering systems can also be a problem; however, in the case of a receiver that uses a channel matched filter, the matched filter will not be matched to the channel between the receiver and the interfering transmitter. This tends to spread out the interference and to give it a Gaussian distribution.

5.6.4 Ipatov Ternary Sequences

As mentioned previously, *m*-sequences have a two-valued periodic autocorrelation function, which is useful for channel sounding and radar or distance measurement applications. Two-valued autocorrelation is the best type of periodic autocorrelation for binary sequences; as a result, these sequences have become known as sequences with ideal periodic autocorrelation properties. Much research has been published relating to the field of sequences with two-valued periodic autocorrelation.

In 1979, Valery Ipatov published a paper [4] detailing a family of ternary sequences with an even more useful property. The periodic autocorrelation function of these Ipatov ternary sequences is a train of equally spaced, equal amplitude impulses. They are thus said to have perfect periodic autocorrelation.

Tom Høholdt and Jørn Justesen [5] extended the original work of Ipatov and found more ways to generate these sequences.

Ipatov sequences have an additional advantage over *m*-sequences in that they exist for many more lengths than do *m*-sequences. All *m*-sequences are of length $2^n - 1$, where n is a positive integer. Ipatov ternary sequences, on the other hand, exist for all of the following lengths: 7, 13, 21, 31, 57, 63, 73, 91, 127, 133, 183, 273, 307,

341, 364, 381, 511, 553, 651, 757, 871, 993, 1,023, 1,057, 1,407, 1,723, and so on. Most of the lengths are of the form $N^2 - N + 1$, but not all numbers of this type are represented; for example, there is no sequence of length 43. They are all generated from cyclic difference sets (see [5] for the method). The sequences are ternary, which means that they have zeros scattered in among the ones and minus ones. This increases their temporal PAR, but, fortunately, most of the sequences have relatively few zeros—for example, the sequences of length 57 (= $8^2 - 7$) have 8 zeros, and the sequences of length 1,723 (= $42^2 - 41$) have 42 zeros. The proportion of zeros for this type of sequence is approximately $1/N$. As the sequences get longer, this proportion vanishes.

An example length-31 Ipatov ternary sequence is

$$+0+0+0+++00+++++++0++$$

The following is a less-common type of Ipatov ternary sequence generated from a larger difference set; it is also length 31:

$$0+0000++0000+00+0++++0+00$$

Another, with an even greater proportion of zeros, is this length-63 sequence:

$$0000+00000000+0000000000+0000+00000000+000000+00+0+++0000$$

5.7 Processing Gain

DS-UWB usually has a processing gain associated with it. In other words, the SNR or the signal-to-noise + interference ratio (SNIR) at the detector is greater than the SNR or SNIR in the channel. In the case of a dense code (e.g., a binary code), this occurs because the signal power in the chips adds together coherently, whereas the noise in the chips adds together noncoherently, giving a net increase of 3 dB in SNR for every doubling of the length of the code. In the case of a sparse code (e.g., a single one in one chip and zeros in the other chips), the noise in the zero-valued chips is ignored, so this also gives a 3 dB-better SNR for every doubling of the length of the code. Another way of looking at this is that the signal is spread over a spectrum that is wider, often a lot wider, than the symbol rate. In the receiver, this signal is then folded back into a bandwidth equal to half the symbol rate. This increases the signal power coherently and the noise noncoherently, which increases the SNR by 3 dB every time the spectrum is folded in two. The amount of processing gain for a real DS-UWB modulation system (as opposed to complex modulation, e.g., BPSK) is two times the ratio of the noise bandwidth to the symbol rate.

$$PG_{real} = 2 \frac{NoiseBandwidth}{SymbolRate}$$

In a conventional direct-sequence system, the bandwidth is constrained to a certain maximum, which limits the chip rate. The symbol rate is then the chip rate divided by the number of chips per symbol. This means that the processing gain is traded off directly against the symbol rate. In a DS-UWB system, however, the signal bandwidth, thus the noise bandwidth, is dependent only on the pulse shape and can be made wider with a consequent increase in processing gain or narrower with a consequent reduction.

5.8 DS-UWB Advantages versus Nonspread Spectrum Methods

5.8.1 High Processing Gain

The spreading of the information out over a wide spectrum gives a large processing gain, which allows signals with a negative SNIR to be recovered.

5.8.2 Immunity to Distortion

The processing gain gives greater immunity to distortion as well as noise. This allows the signal, for example, to be recovered despite a large amount of quantization noise, which in turn allows a relatively small number of bits to be used in the receiver's analog-to-digital converter (ADC).

5.8.3 Wider Bandwidth

The complexity and power consumption of the ADC are exponentially related to the number of ADC bits and linearly related to the sample rate. A small number of ADC bits allows a high receiver sample rate and, therefore, a large instantaneous receiver bandwidth.

5.8.4 Greater Channel Capacity

From Shannon's Law, we know that the channel capacity is proportional to the instantaneous bandwidth and to the log of the SNR plus one.

$$C = W \log_2 (1 + S/N)$$

where C is the channel capacity, W is the instantaneous bandwidth, and S/N is the signal power divided by the noise power.

Some modulation schemes (e.g., Bluetooth) use frequency hopping of a small instantaneous bandwidth, which allows greater instantaneous power to be used while still keeping the average power below the maximum allowed by the spectral regulation authorities. This has the effect of increasing the SNR at the expense of the instantaneous bandwidth. In the case of a frequency-hopping system that uses m hops, the equation above may be rewritten as

$$C_m = \frac{W}{m} \log_2\left(1 + mS/N\right)$$

For m greater than one, C_m is always less than C, so using the wider instantaneous spectrum enables a potentially larger channel capacity.

5.8.5 Fading Immunity

Wireless channels subject to multipath distortion are subject to frequency selective fading. The distribution of the severity of the attenuation at any frequency follows a Rayleigh distribution. At some frequencies, the SNR is very good, and at others it is very poor.

Different modulation schemes have different ways of combating this; for example, discrete multitone (DMT), a relative of orthogonal frequency division multiplexing (OFDM), which is used in digital subscriber line (DSL) modems, combats this using a technique known as bit loading. At frequencies where the SNR is very good, it uses a higher-order modulation and sends a lot of bits per symbol. At frequencies where the SNR is poorer, it sends fewer bits per symbol. At some frequencies, it can decide not to send any information. The bit loading algorithm allows the DSL modem to approach the "water pouring" solution for transmit spectrum allocation, which is known to be the capacity-achieving transmit spectrum. The transmitting modem is able to distribute the information in an optimum way like this because it gets information about the channel from the receiving modem.

Similarly, telephone band modems, like V.34, use a scheme known as precoding. This method modifies the transmitted signal so that the amount of information sent in any part of the band is proportional to the SNR that the receiver sees in that part of the band.

In wireless systems, however, the channel is usually assumed to be nonstationary (i.e., changing very rapidly). This means that the modems cannot know what it will be like from one packet to the next, so techniques like bit loading and precoding cannot be used.

DS-UWB does not spread the information optimally the way DMT does with bit loading. On the other hand, with DS-UWB, each information bit is always spread across the whole bandwidth in use, so high attenuation at a particular frequency is not so damaging. The processing gain of DS-UWB also helps the problem by folding in the highly attenuated regions in the band with other regions that are probably not so highly attenuated.

5.9 Transmitter Structure

Figure 5.7 is a simplified block diagram showing the main elements in an example DS-UWB transmitter.

The transmit data is scrambled to remove any correlation between adjacent bits. These scrambled bits are then coded with a convolutional code. Groups of these coded bits are then taken, and their value is used to select a code to transmit. In a simple BPSK system, the two choices would be a code or its negative. The codes are then fed into the pulse generator, chip by chip. If the chip is positive, a pulse is sent; for zero, no pulse is sent; if the chip is negative, a negative pulse is sent. The pulse generator is designed to send pulses with a spectral shape close to the desired final shape. The resultant train of pulses is sent through a filter that fine-tunes the spectral shape and ensures compliance with local regulations. The signal is then fed into either a switch or a two-wire/four-wire (2W/4W) hybrid, which is in turn connected to the antenna. In many wireless systems, a switch is used here. At any one time, the radio is usually either transmitting or receiving. When it is transmitting, the switch position is set so that the transmitter is connected to the antenna. When the radio is receiving, the switch connects the receiver to the antenna. Because UWB power levels are very low, a switch is usually not required, and a 2W/4W balancing hybrid works well. Like a switch, the 2W/4W balancing hybrid converts the four-wire transmitter and receiver signals into a two-wire signal that can be connected to an antenna. It uses a balancing network or bridge to make sure that not too much transmit power gets from the transmitter across to the receiver.

5.9.1 Transmitter Design Choices

Many trade-offs may be made in the transmitter design. For example, a very poor pulse generator can be used, one that generates lots of disallowed out-of-band

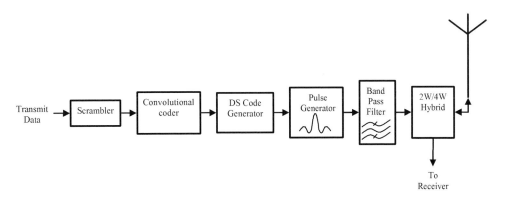

Figure 5.7 Typical DS-UWB transmitter structure.

164 Chapter 5

energy. In this case, however, a good, highly selective transmit filter with good roll-off is required to round off the signal and ensure that it meets regulatory spectral requirements. On the other hand, if the transmit pulse shape is very good to begin with, the transmit filter can be a simple one, which just finishes off the signal.

Another choice is between generating the pulse in the baseband and mixing it up to the required frequency or designing the generator to be able to construct the pulse at the required frequency in the first place.

5.9.2 System Clock Choice

It is important that as many of the system clocks as possible—ideally, that all of the clocks—be related to one system clock. This relationship should be a simple one; for example, a single, commonly available crystal might generate the master clock. This is multiplied up to get what can be a very high frequency internally, and all of the transmitter clocks needed are got by dividing this down by an integer, preferably by a power of two. This can be brought about by careful system design. The frequencies to take care over are, for example,

- Center frequency or carrier frequency of the system
- Chip rate clock
- Bit rate clocks

5.10 Receiver Structure

Figure 5.8 is a simplified block diagram showing the main elements of a example DS-UWB receiver.

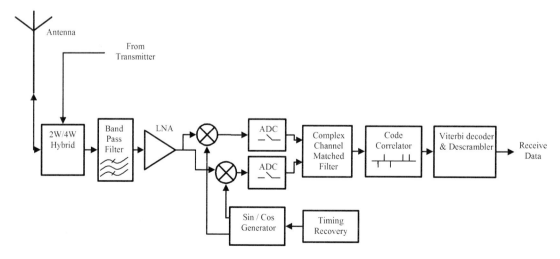

Figure 5.8 Typical DS-UWB receiver structure.

In the example shown, the receive signal goes into the hybrid from the antenna. This is filtered to remove any out-of-band interference and then amplified using a low-noise amplifier. The output is then demodulated by a quadrature demodulator. The demodulator is fed by a local oscillator, which is kept synchronized to the remote transmitter by a timing recovery circuit. The output of the demodulator is fed into two ADCs, one each for the real and imaginary demodulator outputs. These are fed into a complex channel matched filter. The matched filter output is correlated with the code possibilities, and the bits corresponding to the best match are fed into the Viterbi decoder. The decoder output is descrambled to recover the receive data.

5.10.1 Matched Filter Training

As mentioned previously, an ideal DS-UWB receiver has a channel matched filter. This filter is often implemented using a rake receiver. In order to calculate the weighting of the taps in the receiver, the channel impulse response needs to be determined. A training sequence can be used for this purpose, and the ternary Ipatov sequences discussed earlier are ideally suited to this purpose, provided that the sequence used is longer than the channel impulse response.

5.10.2 Receiver Design Choices

As with the transmitter, the receiver designer is faced with choices and trade-offs. The choices include

- Equalizer: A more sophisticated receiver with an equalizer before the correlator would be required for very high rates.
- Number of ADC bits: The receiver can operate with only one bit in the ADC. The more bits that are used, the smaller the loss due to quantization noise, but the more expensive and power hungry the device will be.
- Rake receiver: The number of rake taps and how many bits to use in each needs to be decided. Again, this is a performance versus power and cost trade-off.
- Receive filtering: The amount of rejection in the receive filters needs to be decided. This will depend a lot on the expected environment; for example, a device that will be mounted on the same circuit board as a 5–6 GHz 802.11a transmitter might require a second receive filter after the low noise amplifier (LNA), whereas a device operating in a country that does not allow commercial use of the 5–6 GHz band might need very little rejection for that band.
- Code correlator placement: The channel matched filter and code correlator are two linear operations, and this means that the order can be swapped around. This can result in reduced complexity by taking only symbol-rate input to the channel matched filter, but it can complicate other system components, such as the equalizer.

Channel Model	Mean Range (m)			
	110 Mbps	220 Mbps	500 Mbps	660 Mbps
No multipath	23.4	16.5	8.5	9.1
LOS 1m–4m	17.0	11.7	5.6	5.3
NLOS 1m–4m	14.7	9.8	5.0	4.5
NLOS 4m–10m	14.3	9.3		
NLOS 4m–10m worst-case multipath	14.0	8.8		

Table 5.4 Simulation Results

5.11 Simulation Results

The sets of simulation results shown in Table 5.4 were obtained from a fully impaired Monte Carlo simulation of a DS-UWB communications system. The carrier frequency used was 3.96 GHz, and the chip rate was one-third of this at 1.32 GHz. The system used a rate 1/2 convolutional code with a constraint length of six bits. For the 500 Mbps bit rate, the code was punctured to rate 3/4. The various bit rates were obtained by using direct-sequence codes of different lengths. They varied from length 6 for 110 Mbps to length 1 for 660 Mbps. The receiver configuration was similar to one shown in Figure 5.8 but used a decision feedback equalizer. The matched filter was implemented with a 16 tap rake with taps quantized to three bits. The ADC had three-bit quantization. The simulation was carried out over a perfect channel and four types of multipath channels. The amount of multipath simulated by these channels varied depending on whether the channel was line of sight (LOS) or not (NLOS) and on what range of distances were being simulated.

The results show that very useful distances of between about 4m and 20m can be obtained by a DS-UWB system at very high bit rates. The transmit power used in all cases, approximately –11 dBm, is very low compared to conventional wireless communications systems.

References

1. Proakis, John G. *Digital Communications, 2nd ed.* New York: McGraw-Hill, 1989.
2. Golay, M. J. E. "Sieves for Low Autocorrelation Binary Sequences." *IEEE Transactions on Information Theory* IT-23 (January 1977): 43–51.
3. Rao, K. S., and P. S. Moharir. "Self-co-operative Ternary Pulse-Compression Sequences." *Sādhanā* 21, part 6 (December 1996).
4. Ipatov, V. P. "Ternary Sequences with Ideal Autocorrelation Properties." *Radio Eng. Electron. Phys.* 24 (October 1979): 75–79.
5. Høholdt, Tom, and Jørn Justesen. "Ternary Sequences with Perfect Periodic Autocorrelation." *IEEE Transactions on Information Theory* 29, no. 4 (July 1983): 597–600.

6

Multiband Approach to UWB
by Charles Razzell

6.1 Introduction and Overview

Although many companies have worked on and developed single-band impulse radio technology, many of them have independently arrived at a multiband approach to UWB in order to solve commonly perceived problems. These problems include

- Inflexible spectrum mask because the occupied spectrum is dictated in large part by the pulse-shaping filter and cannot be easily altered
- Inherent implementation difficulties of truly wideband circuit design (with multiple gigahertz of bandwidth) as design of active circuits that perform well over large bandwidths is challenging and usually gives rise to increased cost and power consumption
- High sample rates in digital-to-analog converters (DACs) and analog-to-digital converters (ADCs) associated with transmit and receive signal paths, respectively
- Vulnerability to strong interferers, which may completely capture the receiver amplifiers or ADCs, causing the wanted signal to be blocked from further processing
- Single-band UWB signals not so well suited to low-cost RF–complementary metal-oxide-semiconductor (RF-CMOS) implementations, due principally to the problem of designing on-chip selectivity with relatively poor passive components available in deep submicron CMOS processes

These factors are explained in more detail in the following paragraphs. However, before going further, the fundamental concepts of multiband UWB are described (see Figure 6.1).

As Figure 6.1 shows, multiband UWB signaling is simply the division in the frequency domain of a single UWB signal into multiple sub-bands. These sub-bands may be transmitted in parallel or sequentially and may be received by separate receive paths or one single frequency-agile receiver.

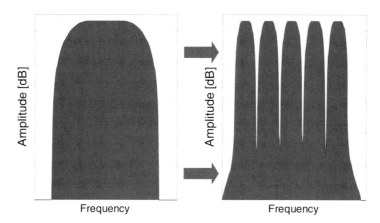

Figure 6.1 Multiband concept for UWB.

Sequential multiband transmissions have two broad classifications:
- Predetermined sequence in the transmitter and known in advance (or estimated in advance) by the receiver. A single receiver may be used, requiring tight synchronization between the local oscillators in the transmitter and receiver, respectively. In this case, the choice of frequency sequences may be used to provide isolation between independent piconets by limiting the number of collisions.
- Variable sequence in the transmitter used to convey modulation information. This technique is usually referred to as Spectral Keying (see Chapter 7). In this case, multiple parallel receivers are needed since the location in the frequency domain of the next impulse cannot be predicted in advance. In this case, the choice of frequency sequences is primarily used to convey information and can typically no longer provide isolation between piconets.

Figure 6.2 Parallel pulsed multiband transmission.

Figure 6.3 Sequential pulsed multiband scheme.

A further classification of multiband schemes needs to be made, namely between pulsed multiband and OFDM multiband approaches, as explained below:

- Pulsed multiband transmissions use a specifically chosen pulse and constant pulse shape to obtain the frequency-domain properties for each sub-band. The spectral shape in each sub-band can be determined by taking the Fourier transform of the selected pulse shape. Typically, phase modulation is applied to convey information. Receiver detection schemes applicable to single-band UWB pulses can be applied, although certain limitations arise from the short dwell time on each sub-band.
- OFDM can be applied in each sub-band. The occupied bandwidth and spectral shape are largely defined by the inverse Fourier transform applied in the transmitter. This technique can be considered as further frequency division of each sub-

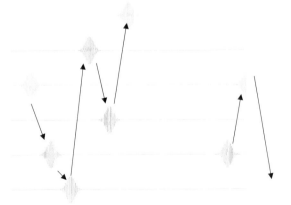

Figure 6.4 Sequential multiband with sequence keying.

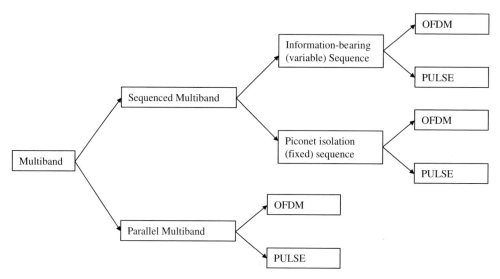

Figure 6.5 General classification of multiband UWB transmission schemes.

band of the multiband scheme into a further *parallel multiband* scheme providing a much finer degree of granularity in the frequency domain.

6.2 Detailed Motivation for Multiband UWB Transmission Schemes

6.2.1 Control of Spectrum Mask

UWB transmissions are limited by their peak power spectral density in the frequency domain in the direction of maximum radiation. Thus, to maximize average power in relation to the legislated peak power, power spectral density should be constant with respect to frequency (white) and with respect to direction (omnidirectional radiation pattern). Control of passband flatness of the power spectrum is thus critically linked to obtaining the optimum average power within the constraints set by the available peak power limits.

Consider the case of a single UWB band transmission scheme where the pulse shape is used to define the frequency-domain properties of the signal. Let's assume that it is desired to occupy, nominally, 1.5 GHz of spectrum with maximal flatness and to avoid as far as possible emissions outside the intentional 1.5 GHz portion of occupied spectrum. In particular, it will be necessary to meet FCC and other regulatory jurisdictions' requirements for low emitted energy across bands occupied by cellular, GPS, and other sensitive services.

Now, if we want to control the spectrum of the 1.5 GHz-wide UWB signal by means of digital pulse shaping, we will be forced to use a sampling rate of at least

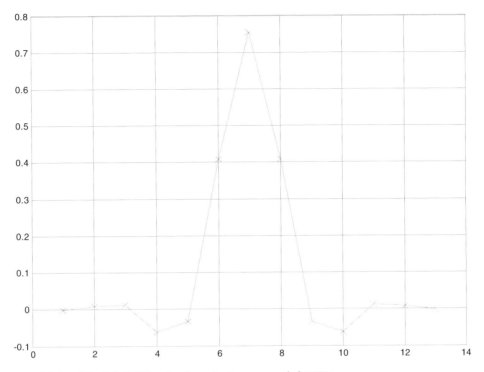

Figure 6.6 Possible digital FIR pulse shape for 2x oversampled UWB system.

twice the Nyquist sampling rate (i.e., 1.5 GHz in each of the two quadrature baseband channels). The number of required taps is unlikely to be much below 13 in order to obtain a flat passband with sufficient roll-off.

For example, the finite impulse response (FIR) filter shown in Figure 6.6 was designed to validate the required number of taps. The delay of the filter is six samples, or 2 ns.

The requirement to suppress aliases, additional analog filtering is required (see Figure 6.7). Considering the baseband implementation of this analog filtering, we see that a lowpass corner frequency of 1 GHz is suitable, which should have at least 40 dB of attenuation at 2 GHz. This implies a fourth-order analog filter based on an elliptic filter prototype (assuming 0.1 dB passband ripple).

This simple design example may not be optimum. Nevertheless, it does illustrate the following points:

- Although the design has attempted to use the lowest oversampling rate and as few filter coefficients as possible, its arithmetic complexity is still quite daunting, requiring a pair of filters each employing 13 multiply-accumulate operations per clock cycle at a 1.5 GHz clock rate, resulting in a total of 39 billion multiply-accumulate operations per second! (The number of multipliers can, of course, be reduced by a factor of 2 if filter symmetry is exploited.)

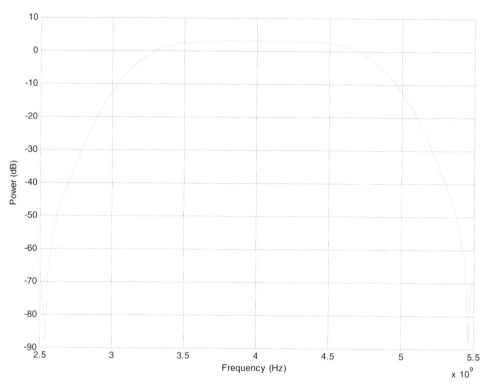

Figure 6.7 Power spectrum resulting from pulse shape described above (aliases not shown).

- Due to the low oversampling rate used, a significant amount of analog filtering is still required to eliminate aliases (e.g., a fourth-order elliptic). This is not trivial to implement either.

In view of these factors, single-band UWB system designers have often resorted to purely analog filtering for pulse shaping. This approach certainly eliminates the computational-complexity issue in the transmitter but also reduces the available accuracy and repeatability of the transmitted pulse shape, with a consequent lack of control over transmitter spectrum flatness and out-of-band attenuation.

By contrast, a multiband approach in the transmitter allows digital filtering for transmitter pulse shaping to be more readily applied. Typically, the 1.5 GHz-wide signal would be split into three bands of 500 MHz each, and only one band need be energized at a time, resulting in two filters of 500 MHz sample rate each for the assumed 13-tap filter. This is still a high computational load, but a much more manageable one, especially when certain optimization techniques are considered. Some relevant techniques to consider are

- Folding the filter to exploit coefficient symmetry
- Redesign using a half-band approach to introduce zero-valued coefficients

- Breaking the filter into N parallel chains, each operating at $1/N$ of the full clock frequency.

Furthermore, the multiband approach also allows the transmit power in each sub-band to be independently managed, providing a means to compensate at a coarse level for frequency dependent attenuation that may arise due to parasitic circuit elements and nonideal antenna frequency response.

To summarize, digital filtering and pulse shaping are much more feasible using a divide-and-conquer approach and reducing the maximum bandwidth to a number in the region of 500 MHz. The precision and repeatability associated with digital filtering are highly desirable in the context of generating UWB signals, especially considering the need for well-controlled spectral flatness in the passband in order to optimize the available power and the resulting link margin. The ability to tune transmitter power in each sub-band further aids in this objective.

6.2.2 Receiver Sampling Rate Issues

The requirement for FIR pulse shaping in the transmitter is mirrored by the requirement for a matched filter in the receiver. Due to the highly dispersive nature of the UWB channel, the filter in the receiver may not be limited to matching the transmitter pulse shape but must be matched to the channel in the form of a channel-matched filter or rake receiver structure. Thus, it should be immediately obvious that a fixed analog filter is not a realistic option and that digital FIR filters with variable tap coefficients are needed.

The requirements of the receiver FIR filtering are different from those of the transmitter. The length of the receiver filter in time should be approximately equivalent to the duration of the channel impulse response. The sampling rate must be greater than or equal to the bandwidth of the transmitted waveform to satisfy Nyquist requirements. (As a rule of thumb, the duration of the significant energy portion of the channel impulse response for an indoor UWB channel over a 10m range is typically between 40 and 60ns.) In general, we can write

$$N_t = \tau_{ch} \times F_S$$

where N_t is the number of filter taps required, τ_{ch} is the excess delay of the channel impulse response based on a particular energy threshold, and F_S is the sampling rate. However, the rate of multiply-accumulate operations required to operation the filter in real time is given by

$$R_{MAC} = N_t \times F_S = \tau_{ch} \times F_S^2$$

The appearance of the factor F_S^2 in the above equation provides a very powerful motivation for multiband approach. For example, reducing the bandwidth and, thus, the value of F_S by a factor of 3 by means of the multiband approach reduces the required rate of multiply-accumulate operations by a factor of 9. Considering a value of τ_{ch} = 40 ns, illustrative numbers are as follows:

- Multiband (F_S = 500 MHz), 10 taps, R_{MAC} = 10×10^9 operations per second
- Single band (F_S = 1.5 GHz), 30 taps, R_{MAC} = 90×10^9 operations per second

This is not meant to imply that one approach is possible while the other is infeasible but rather that when considering the need for low-power consumption, the multiband approach has a clear advantage, even when only three sub-bands are considered. The potency of this argument increases as the number of sub-bands to be used increases.

6.2.3 Active Circuit Bandwidth and Power Consumption

In UWB receivers for sequential multiband modulation schemes, the first down-mixing step results in a bandwidth reduction by a factor equal to the number of sub-bands employed. This is illustrated in Figure 6.8. The section to the left must be capable of passing the full UWB bandwidth signal with high linearity. The section to the right benefits from a bandwidth reduction by a factor equal to the number of sub-bands employed. The mixing stage in the case of the multiband scheme can be seen as a kind of correlator in which the agile local oscillator is synchronized to the one in the transmitter and has similar attributes of selectivity and bandwidth reduction.

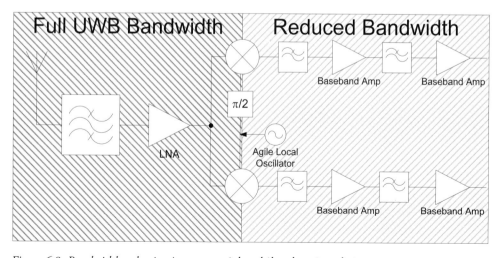

Figure 6.8 Bandwidth reduction in a sequential multiband receiver chain.

This is not true of single-band modulation schemes, where the local oscillator is held at a constant frequency and the action of the first down-mixing step is simply one of frequency translation (usually to a center frequency of dc).

The consequence of this is that on-chip channel selectivity at baseband is much more feasible in the multiband case, where the lowpass corner may be 250 MHz (for a total intermidiate frequency [IF] bandwidth of 500 MHz). For the single-band case, the equivalent lowpass corner would have to be 750 MHz, which means that active on-chip filtering techniques are difficult, if not impossible, to apply. Similarly, the current consumption required to obtain linear operation and high dynamic range up to 250 MHz is much less than that required to reach 750 MHz.

Similar arguments apply to a direct up-conversion transmitter for similar reasons. Once again, the implementation difficulty is reduced for the analog transmitter blocks until the up-conversion with the agile local oscillator spreads the bandwidth to its final value.

6.2.4 ADC and DAC Sampling Rates

By extension of the discussion above, it will be readily apparent that the sampling rates required in the mixed signal components will obtain a reduction by a factor equal to the number of sub-bands employed. This is significant when one considers the digitization of UWB signals a challenge and when one takes into account the need to reduce power consumption to levels sustainable in portable battery-operated equipment. For a multiband system, where each sub-band may typically be 500 MHz wide, the sampling rate of the two required ADC converters needs to be at least 500 MHz to satisfy the Nyquist sampling theorem. Typically, between 4 and 6 bits of precision are used. A similar single-band system occupying 1.5 GHz of spectrum would require a sampling rate of at least 1.5 GHz. However, if a correlator or despreader is to be used as part of the receiver's digital signal processing, the required ADC precision may be lower (e.g., 3 bits), which may wholly or partially compensate for the extra power consumption required by the higher sampling rate. The difference in required precision stems from the fact that in the sequential multiband scheme, despreading is done in the analog domain. The ADCs are effectively operating on a partially despread signal. In the case of a single-band scheme, the despreading or correlation is done entirely in the digital domain. This despreading or correlation clearly adds precision to the resultant signal since summation of several samples with uncorrelated quantization noise must increase the signal-to-quantization noise ratio.[1]

---

[1] A similar exchange of precision for sampling rate can be observed in the decimation stages of Sigma-Delta ADC. This has led some researchers to consider using a 1 bit ADC with an ultra-high sampling rate for UWB receiver applications.

6.2.5 Vulnerability to Strong Interferers

The multiband approach has advantages with respect both to in-band and out-of-band interference rejection. The main advantage with respect to out-of-band interferers is that the required baseband filtering shape factor is eased considerably in the case of the multiband approach for a given frequency offset between the interferer and the nearest passband edge of the UWB system. For example, consider a 3.2–4.8 GHz single-band UWB system that must reject an interferer at 5 GHz. The UWB system has a 4 GHz center frequency, with an upper passband edge at 4.8 GHz, but a rejection requirement at 5 GHz. The interferer to be rejected is only 0.32 octaves above the passband edge since $\log_2((5{,}000 - 4{,}000)/(4{,}800 - 4{,}000)) \approx 0.32$. Now, consider a multiband UWB system with three sub-bands of 528 MHz bandwidth centered at frequencies of 3,480 MHz, 4,008 MHz, and 4,536 MHz. In this case, rejection of the 5 GHz interference signal primarily concerns the upper sub-band centered at 4,536 MHz and having an upper passband edge of 4.8 GHz. The interferer to be rejected is now $\log_2((5{,}000 - 4{,}536)/(4{,}800 - 4{,}536)) \approx 0.814$ octaves above the passband edge, even though the absolute offset remains 200 MHz in each case. We conclude that the filter order required for the single-band case would have to be approximately 2.5 times greater to achieve the same interference rejection as the multiband case. This is relevant since UWB transceivers must be designed to coexist at close proximity with relatively high power emissions from commonly used unlicensed devices in the 2.4 GHz Industrial, Scientific, Medical (ISM) band and the 5 GHz unlicensed national information infrastructure (UNII) radio band.

We now turn our attention to in-band interferers (i.e., those that overlap the transmission band used by the UWB system to be designed).

One significant difficulty with UWB receivers employing low-precision ADC conversion schemes is dealing simultaneously with a desired signal at threshold sensitivity and a strong, in-band interferer. Although a well-designed automatic gain control (AGC) circuit can provide for a wide dynamic range of wanted signals, AGC cannot provide any help in the situation where a strong in-band interferer is present at the same time as a wanted signal possibly 60 dB lower in amplitude. The response of a pair of 3 bit ADCs to such a situation is likely to be such that the wanted signal is either below the quantization noise floor, thus blocked from further processing, or the unwanted signal saturates the ADCs, also blocking the signal from further processing. Thus, protection from strong in-band interferers under these conditions requires analog filters (band-reject filters or traps). This approach is best suited to situations where the location of the interference is known in advance since providing an analog tunable notch with an ultrawide tuning range is challenging, as is the ability to detect dynamically where the notch is required.

Multiband UWB receivers offer a redundant approach to reception over the full UWB spectrum. It is likely that a strong source of interference may only impact one of the sub-bands. Two methods are available for exploiting the available frequency diversity:

1. The receiver alone adapts by inserting erasures for the symbols to be received in the interference impacted sub-band
2. The transmitter and receiver negotiate to adopt a smaller or different set of sub-bands, avoiding the useof the sub-band that has been interfered with heavily.

In the case of receiver-based adaptation, very little needs to be done. Even if the source of interference is continuously present, it will only be present at the ADCs during the time that the local oscillator is tuned to that sub-band. Provided that the ADCs can recover quickly from the saturation condition (or, otherwise, that the AGC algorithms and hardware are sufficiently agile), only $1/N$ of the information to the receiver will be blocked, where N is the number of sub-bands employed. With sufficient redundancy in the form of forward error correction coding and repetition coding, the impact of losing an entire sub-band may be limited to a reduction in the highest available data rate or in range, but it is unlikely that the entire communications link will be cut, which remains a risk for the single-band approach to UWB.

The second approach requires coordination between the transmitter and receiver and, thus, incurs a protocol overhead. However, the advantage of this approach is that the same Eb/No can be maintained for the same transmitted data rate.

Figures 6.9 and 6.10 illustrate a possible response to a strong interferer in one sub-band of a three-sub-band multiband UWB system.

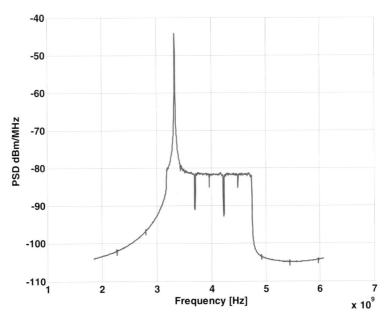

Figure 6.9 Multiband signal received with a strong in-band interferer.

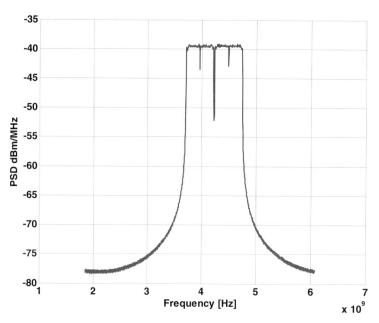

Figure 6.10 Transmitted multiband with one sub-band removed after detection of a strong interferer.

6.3 Multipath Energy Collection in Sequenced Multiband Receivers

Indoor UWB channel models are highly dispersive relative to the pulse widths typically considered. One issue with sequential multiband receivers is the requirement to dwell on a band long enough to allow the dispersed energy to be collected before moving on to the next sub-band. This section explores this problem and some potential solutions.

6.3.1 Example Case Study

Consider the case of a sequential pulsed multiband scheme as follows:

- 264 MHz pulse repetition frequency (PRF)
- Binary phase shift keying/quadrature phase shift keying (BPSK/QPSK)
- Rate 1/2 convolutional code
- 264 Mbps maximum data rate
- Seven bands spaced at 440 MHz

The nominal dwell time on each band is 1,000/264 = 3.79 ns. However, the channel excess delay may easily exceed 40 ns. So, we can see that significant performance loss can be expected due to the amount of energy that is discarded when the local oscillator switches to a new frequency. Lower data rates can make use of a reduced PRF (e.g., a double dwell period) on each band to increase energy collection, but this is only applicable when the supported data rate is halved to 132 Mbps and below.

Clearly, increased dwell time (in order to collect the dispersed multipath energy) conflicts with the higher PRF values that may be needed to support higher data rates. One method to achieve higher data rates is to consider higher-order modulation schemes. However, since modulation orders above 4 quadrature amplitude modulation (QAM) are not desired in the case of a severely power-limited system like UWB, we reject this option and turn our attention to increasing the number of parallel receivers.

Figure 6.11 illustrates one such scheme. Although the scheme operates over seven bands, it is not necessary to have seven parallel receivers; in fact, only two parallel receivers are employed. The dwell time is thus increased by a factor of 2. The local oscillators for all seven bands are generated concurrently, and a switching mechanism is employed to provide the correct sequencing of local oscillator frequencies to the down-conversion mixers. Due to the extended dwell times, the local oscillators must provide 50 percent overlap, as illustrated in Figure 6.12.

Considering the two methods of increasing the dwell time on each band, the number of possible RAKE fingers may be 1, 2, or 4 as shown in Table 6.1. Clearly,

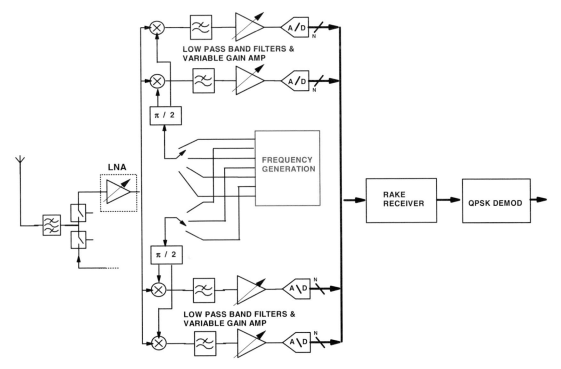

Figure 6.11 Direct conversion receiver architecture with two parallel branches to lengthen dwell time.

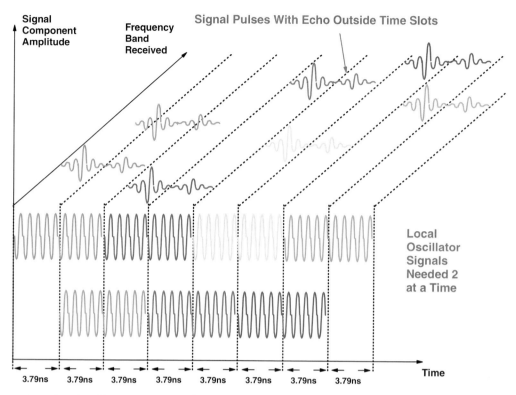

Figure 6.12 Local oscillator requirements for double-dwell receiver architecture.

bit error rate (BER) performance will be highly dependent on this number, and simulation results bear this out. Figure 6.13 shows the performance impact of the number of fingers employed for the severe case of channel model 4. The best performance is obtained with schemes using four fingers, requiring two parallel receivers, with similar performance being obtained for BPSK or QPSK schemes. However, these schemes are limited to 132 Mbps in the case of QPSK or 64 Mbps in the case of BPSK. If 264 Mbps is required, a full PRF scheme using two fingers is the best possible scheme, and the associated performance loss is found to be approximately 1 dB.

The best performing fully serial scheme is the half-PRF scheme using QPSK and two RAKE fingers, but this only supports 132 Mbps and has a performance loss of 8 dB compared to the four-finger scheme. There remaining serial receiver schemes

	Fully Serial Receiver	Two Parallel Receivers
Half PRF (132 Mbps)	2	4
Full PRF (264 Mbps)	1	2

Table 6.1 Maximum Number of RAKE Fingers as a Function of PRF and Receiver Parallelism

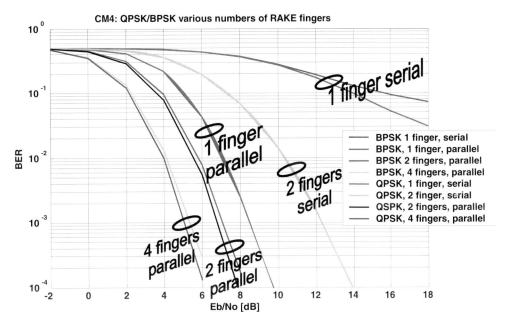

Figure 6.13 BER impact of number of receiver fingers in parallel and receiver configurations.

with only one finger do not have sufficient performance to be viable with this channel model.

6.3.2 Example Case Study Conclusion

Through one example case study, we have observed that a sequential multiband UWB scheme requires long dwell times on each sub-band to collect sufficient multipath energy for good performance. We have also noted that this requirement conflicts with the desire to increase data rates, assuming that only one or two bits are sent with each UWB pulse. One possible solution using receiver parallelism has been detailed here. However, the cost of implementing receiver parallelism is such that many UWB system architects have preferred to consider other approaches, most notably OFDM with its efficient multipath energy collection and high information density (typically 200 bits per OFDM symbol).

6.4 Local Oscillator Generation Aspects

6.4.1 Problem and Candidate Solutions

Many experienced radio engineers will have a strong negative reaction to the prospect of having to hop between widely separated radio channels in a couple of nanoseconds or less. This is understandable since most frequency synthesizers are designed to switch frequencies in milliseconds, perhaps even in hundreds of microseconds, but never in nanoseconds. It quickly becomes apparent that compared to the traditional analog phase locked loop (PLL) and voltage controlled oscillator (VCO), a fundamentally different approach to fast frequency switching is needed.

Two techniques are generally considered where very fast frequency switching is needed: direct digital synthesis and parallel frequency generation.

A typical direct digital synthesis scheme consists of a phase accumulator with a binary adder in the feedback loop. The frequency is directly determined by the phase increment, which is one input to the binary adder. The output of the phase accumulator is fed to a sine or cosine lookup table, and these digital values are fed to a DAC to provide the variable frequency output. Such a scheme has the desired merit that frequency changes are virtually instantaneous once the phase increment value is changed. However, the rate at which phase accumulations must take place to synthesize frequencies in the gigahertz range (not to mention the digital-to-analog conversions) may stretch the limits of feasibility with today's CMOS processes and, even if feasible, will likely result in disproportionately high power consumption.

Two approaches to parallel frequency generation are described here, the first using multiple oscillators with one oscillator dedicated to each sub-band and the second using single side band (SSB) mixers to create frequency shifts from a pair of oscillators.

6.4.2 Multiple Parallel Oscillators

In this approach, each sub-band has a dedicated oscillator. In order to make this concept attractive for a monolithic integrated design, on-chip ring oscillators may be considered. Such oscillators have the advantage of requiring relatively little silicon real estate due to the absence of inductors. Another advantage is that if the ring oscillator consists of four inverters (or a multiple thereof), quadrature outputs can be derived directly from an appropriately selected pair of nodes in the ring. Since each ring oscillator must be phase locked to a common crystal-derived frequency reference, the ring oscillators must be voltage controlled. Furthermore, phase noise is a prime consideration for such a system and requires careful design [1]. Care must also be taken with layout, output buffering, and power supplies to prevent unintentional coupling between the ring oscillators.

A disadvantage of this approach is that each ring oscillator must be kept powered on at all times during active receive or transmit operations with its associated divider and PLL circuit. This can result in excessive power consumption, depending on the number of sub-bands employed. For UWB systems employing only three sub-bands, the power consumption overhead may be considered tolerable.

6.4.3 Frequency Generation Using SSB Mixing Principles

Another approach to frequency generation is to start with a fixed frequency oscillator locked to one of the desired sub-band frequencies. An SSB mixer performs frequency shifts in a positive or negative direction, with the magnitude of the shift being determined by a second lower-frequency oscillator. Note that if dc is selected instead of the lower-frequency oscillator, zero frequency shift is obtained. Inverting or not inverting one of the quadrature components of the lower-frequency oscillator determines the sign of the frequency shift.

It should be readily apparent that the frequency-switching time is determined by the speed with which the integrated selector switch can fully change between the three possible states corresponding to negative, zero-valued, and positive frequency shifts. This allows switching times on the order of nanoseconds, as required.

A disadvantage of this approach is the generation of spurious tones, especially odd harmonics of the lower-frequency oscillator used for switching. This may result in the need for on-chip filtering to suppress them.

Extensions to this scheme may be considered to generate more than three frequencies, for example, by cascading more SSB mixer stages to allow additional frequency shifts. Alternatively, the frequency generator used for shifting may be implemented using a cascaded divider network, allowing the frequency-shift value to be doubled or halved according to the selected node in the divider chain. Increased

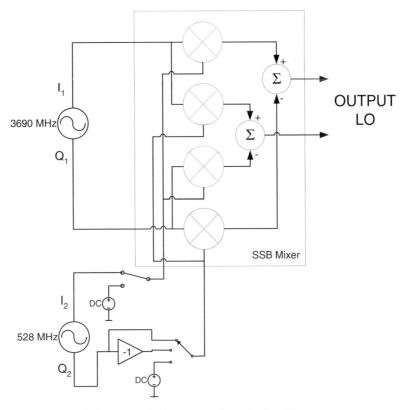

Figure 6.14 Frequency shifting using SSB mixers and a pair of oscillators.

care with spur suppression will be needed as the topology and number of mixers are increased.

6.5 Regulatory Implications of Multiband UWB Transmissions

The U.S. FCC was the first major regulator to provide a legislative framework for UWB transmissions. The multiband approach to UWB was not widely known or considered at the time that the FCC published its first Report & Order relating to UWB in February 2002.

6.5.1 Minimum Bandwidth Requirements

The FCC has determined that the minimum UWB bandwidth shall be 500 MHz or 20 percent of the carrier frequency (whichever is smaller). An important corollary is that this minimum bandwidth shall be maintained at all times. Thus, it is not sufficient to meet the 500 MHz requirement by hopping or frequency sweeping. This restriction is understood to mean that, for multiband systems, each sub-band must by itself meet the minimum bandwidth requirements of the FCC for UWB systems by having a 10 dB bandwidth of at least 500 MHz.

6.5.2 Mean Power Requirements

The original text of FCC Report & Order 02-48 and its subsequent revisions have maintained that measurements of frequency-hopped and frequency-swept systems must be carried out with the frequency sweeping or hopping turned off. Although the allowed transmit power of a UWB system is usually proportional to its occupied bandwidth in normal operational mode, this rule effectively limits the average power of a multiband UWB system to a value that corresponds to the width one of its sub-bands. This has been a very controversial issue due to the possible implication that the transmit power may have to be reduced by a factor of N, where N is the number of sub-bands employed. In the United States, this topic has been the subject of a successful petition for waiver by multiband OFDM proponents.

Beyond the outcome of the petition for waiver, it is important to understand some of the pros and cons of allowing a frequency-hopped system to be measured under normal operating conditions. An understanding of these issues should inform future system designs employing multiband techniques.

6.5.3 Impact of Multiband UWB Transmissions on Victim Receivers

Typically, existing services have a much narrower bandwidth (e.g., < 30 MHz) than one sub-band of a multiband UWB system. In cases where a victim service overlaps with one of the UWB sub-bands, the victim receiver will experience higher interfer-

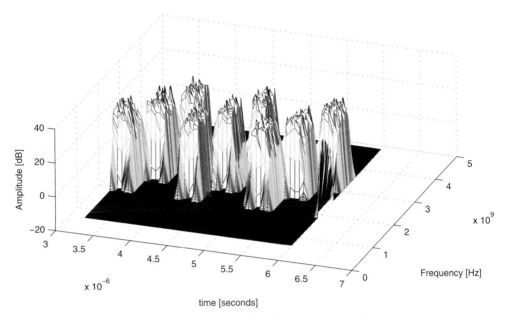

Figure 6.15 3-D spectrogram of an N = 3 multiband transmission.

ence power when that sub-band is active and negligible interference at other times. Hence, from the point of view of the victim receiver, multiband UWB interference has a duty cycle of 1:N, where N is the number of sub-bands employed. However, the average power spectral density seen by the victim receiver density is maintained at or below the FCC's limit value of −41.3 dBm/MHz. It follows that the interference power received by the victim *during the time that the overlapping sub-band is active* will be increased by a factor of N, compared to the case of a stationary waveform. This can be visualized by referring to the example in Figure 6.15.

Figure 6.16 shows the same waveform in "plan" view; in addition, a pair of parallel lines is used to represent the bandwidth limits of a potential victim receiver.

Critics of multiband UWB technology have argued that the interference potential has been increased by a value of $10\log_{10}(N)$ dB due to the higher peak power consistent with the duty cycle of the UWB interference. Consequently, it has been argued that the power of multiband UWB transmissions should be reduced by a factor N to compensate. On deeper examination, the situation is naturally mitigated by several factors:

1. The above discussion assumes that the only source of interference is the UWB signal and that the victim receiver is operating at far above its own thermal noise floor. In fact, the desensitization caused by a multiband UWB interferer is significantly less than that for victim receivers that are operating only a few decibels above their sensitivity threshold. The desensitization can be said to approach $10\log_{10}(N)$ dB asymptotically only at very high levels of wanted signal, but in

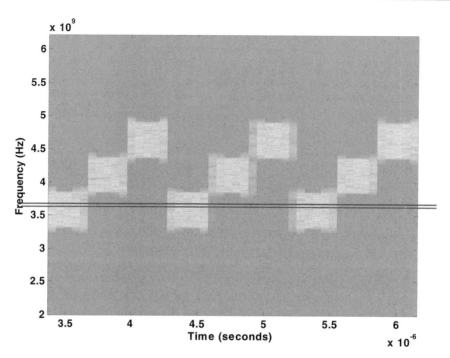

Figure 6.16 Colormap spectrogram of N = 3 multiband transmission.

these cases, it is unlikely that the UWB signal will be strong enough to bring the I/N ratio into the region that may cause a problem.

2. Depending on the pulse rate and the victim receiver's bandwidth, the victim receiver may not be able to track the peaks and troughs of the duty cycle imposed by the multiband UWB waveform, which is a good thing. In cases where the bandwidth is narrow compared to the PRF on each sub-band, the receiver channel filter effectively averages the UWB interference waveform, and the impact is closer to that of a stationary waveform of the same average power. Consider, for example, the case study discussed earlier in Section 6.3.1, where a seven-band hopping system had a 264 MHz PRF, resulting in a PRF of 264/7 = 37.7 MHz per band. In this case, typical victim receivers below 30 MHz bandwidth would benefit by the inherent smearing or averaging caused by their inherent channel selectivity.

3. Restraint in the choice of the value of N can and should be exercised by the system designer, especially in systems that use a relatively low PRF per band. UWB regulations in the United States have allowed for significant latitude in peak-to-mean ratio, for example, to accommodate pulse waveforms. The peak limit for UWB set by the FCC is 0 dBm measured in a 50 MHz bandwidth. The mean value is limited to –24.25 dBm in the same bandwidth, thus allowing for a theoretical peak-to-mean ratio of up to 24 dB. By keeping the value of N relatively

low, especially where low PRF systems are contemplated, it is possible to work well within this peak-to-mean ratio limit and, thus, to limit the interference potential to values significantly lower than those anticipated by the original FCC Report & Order, such that widespread deployment of multiband UWB devices can have a negligible impact on existing services.

In order to quantify the first item in the above discussion, we can directly derive the average BER for a QPSK victim service under the influence of interference from a multiband UWB system with a duty cycle of 1:N

Given the analytical expression for BER of a QPSK modulation scheme, that is,

$$BER = 0.5 \mathrm{erfc}\left(\sqrt{E_b/N_0}\right)$$

we can easily show that when the presence of the noiselike interference is governed by a regular duty cycle, the BER equation is modified as follows:

$$BER' = (0.5/d) \cdot \mathrm{erfc}\left(\sqrt{1/N}\sqrt{E_b/I_0}\right)$$

We consider the case where thermal noise and background interference is negligible to be a special case of academic interest only. Thus, we introduce a constant background thermal noise density N_0 and consider the impact of an interference source of average spectral density I_0, which is subject to on/off keying according to a regular duty cycle factor d. We therefore obtain

$$BER = 0.5 \cdot \mathrm{erfc}\left(\sqrt{E_b/(N_0 + N \cdot I_0)}\right)$$

when the UWB/noise burst is keyed "on" and

$$BER = 0.5 \cdot \mathrm{erfc}\left(\sqrt{E_b/N_0}\right)$$

at all other times. Combining these, we obtain an average BER as

$$\overline{BER} = 0.5\left(\mathrm{erfc}\left(\sqrt{E_b/(N_0 + N \cdot I_0)}\right)/N + (N-1)\mathrm{erfc}\left(\sqrt{E_b/N_0}\right)/N\right)$$

Figure 6.17 illustrates this relationship.

In the case where a continuous noiselike UWB interferer would be tolerated at −10 dB, the multiband system with $N = 3$ has a reduced tolerance of −11.8 dB, a reduction of 1.8 dB. In the case where the threshold required BER is assumed to be

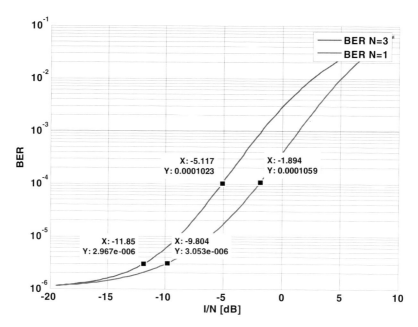

Figure 6.17 Average uncoded QPSK BER as a function of the I/N ratio.

10^{-4}, the tolerable I/N ratio is −1.9 dB for the continuous noiselike UWB interferer and −5.1 dB for the multiband system with $N = 3$, a reduction of 3.2 dB in the tolerable interference level. In each case, the interference tolerance degradation is less than the asymptotic value of $10\log_{10}(N) = 4.77$ dB. Furthermore, this simple calculation does not take in account the beneficial impact of the victim receiver's channel filter as discussed above; nor does it take into account the impact of the forward error correction scheme of the victim service.

For a detailed and thorough investigation of these interactions, please see the NTIA report 05-429 [2].

6.6 Conclusion

This chapter has described the concept and classification of multiband UWB systems, along with the general motivation for employing this technology. The issue of multipath energy collection, a particular pitfall with high-PRF multiband schemes, has been described with reference to a case study. We have also considered the primary techniques usually considered for generating the fast sequences of local oscillator (LO) signals needed for transmission and reception of multiband UWB signals. Finally, an overview of the regulatory issues associated with multiband UWB has been given. The above material is intended to present the pros and cons of the multiband approach so that future system designers and those considering technology options in UWB can make well-informed decisions.

References

1. Hajimiri, A., and T. H. Lee. "Phase Noise in Multigigahertz CMOS Ring Oscillators." *Custom Integrated Circuits Conference Digest* (May 1998): 49–52.

2. Cotton, M., R. Achatz, J. Wepman, and P. Runkle. "Interference Potential of Ultrawideband Signals Part 2: Measurement of Gated-Noise Interference to C-Band Satellite Digital Television Receivers." NTIA Report 05-429, August 2005.

3. Roovers, R., D. M. W. Leenaerts, J. Bergervoet, K. S. Harish, R. C. H. van de Beek, G. van der Weide, H. Waite, Y. Zhang, S. Aggarwal, C. Razzell. "An Interference-Robust Receiver for Ultra-Wideband Radio in SiGe BiCMOS Technology." *IEEE Journal of Solid-State Circuits* 40, issue 12 (December 2005): 2563–2572.

7

Spectral Keying™: A Novel Modulation Scheme for UWB Systems
by Naiel K. Askar, Susan C. Lin, and David S. Furuno

7.1 Background

In 2001, General Atomics began investigating the use of multiband modulation schemes in anticipation of the UWB Report & Order from the FCC. These modulation schemes break up the UWB spectrum into smaller sub-bands, such as those shown in Figure 7.1. Prior to this, the team had been manipulating the polarity, amplitude, and position of UWB pulses, but research quickly focused on multiband approaches because of their obvious advantages. By controlling the frequency content of each UWB pulse, another parameter was available for the optimization of the UWB waveform [1–3].

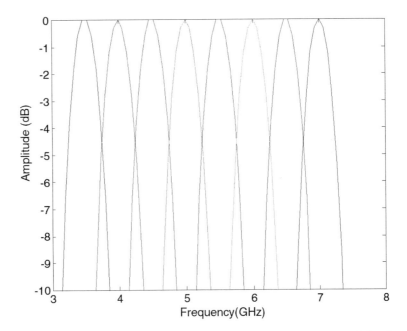

Figure 7.1 A multiband waveform breaks up the UWB spectrum into multiple sub-bands.

The advantages of multiband waveforms have already been well documented [1–3]. In summary, the use of multibands provides an elegant means to deal with interference sources within the UWB band, scalability to lower or higher data rates, and flexibility in the international regulatory environment.

General Atomics' multiband radios employ high-order symbols that use multiple sub-bands to construct each symbol. The symbols are made from short pulses, similar to impulse radios; however, no attempt is made to make the pulses very narrow and impulselike. Instead, the pulses are generated in multiple sub-bands. This has the advantage of providing high data content per symbol, reducing the pulse repetition frequency (PRF) required for a given data rate. By using this approach, it became possible to avoid the intersymbol interference (ISI) caused by delay spread in the propagation channel without additional processing. The result was a simple, high-performance implementation architecture that can be realized in custom integrated circuits (ICs) at low cost and power. Because of the reliance on spectral sub-bands, this modulation approach was referred to as spectral keying (SK).

In this chapter, SK modulation is described in Section 7.2. Section 7.3 gives the performance predictions for these waveforms. Section 7.4 discusses the SK transmitter, and Section 7.5 presents an optimal receiver and receiver architectures.

7.2 Description of SK

7.2.1 Modulation Waveform

In SK, the UWB spectrum is divided into multiple sub-bands. Each sub-band has at least 500 MHz of bandwidth to ensure compliance with FCC rules. SK transmission consists of two or more pulses, each in a different sub-band. Figure 7.2 shows two SK symbols, each made up of four pulses, where each pulse is in a different sub-band, and the corresponding frequency spectrum. Symbols are separated by a guard period, which is needed to mitigate ISI.

In SK modulation, the order of the sub-bands defines the symbol. For example, in Figure 7.2, the first symbol has the sequence $\{F_1, F_2, F_3, F_4\}$, while the second symbol has the sequence $\{F_2, F_1, F_3, F_4\}$. Hence, a two-band system has 2 possible symbols, a three-band system has 6 possible symbols, and a four-band system has 24 possible symbols. The number of possible symbols increases with the factorial of the number of bands. The number of available data bits per symbol is given by $\log_2$ of the number of available symbols, giving 1, 2.6, and 4.6 bits for SK system with two, three, and four sub-bands, respectively. In addition, one can encode information in the phase of each subpulse. This further increases the information content within the symbol.

Spectral Keying™: A Novel Modulation Scheme for UWB Systems   193

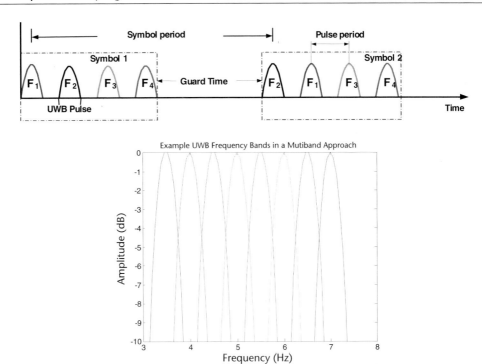

Figure 7.2 Time-domain waveforms of four UWB pulses, together with the corresponding frequency-domain spectrum.

Figure 7.3 is a general representation of the SK symbol where M is the number of sub-bands and T is the number of time slots for the UWB pulses. The symbol X is represented by the matrix, with a 0 signifying no sent pulse, ±1 indicating the phase of the pulse in a BPSK system, ±i indicating quadrature phase for QPSK.

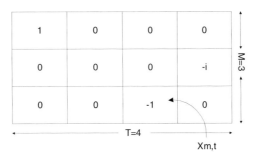

Figure 7.3 Matrix representation of a general SK symbol.

If B is defined as the number of nonzero entries, and P represents the number of phase bits, then the number of encodable bits is defined by

$$N(M,T,B,P) = \binom{M}{B}\binom{T}{B} B! 2^{BP}$$

where

$$\binom{M}{B} = \frac{M!}{M!(M-B)!}$$

7.2.2 Modulation Features

SK is a multiband UWB modulation, and, as such, it is a flexible technology that can easily deal with interference sources or international regulatory restrictions by altering the number and frequencies of sub-bands. In addition, it is scalable to lower or higher data rates by tailoring the number of bands and phase bits employed.

However, the primary advantage of SK results from the high information content of each symbol. A fundamental issue in UWB communications system design is the ISI caused by delay spread in the propagation channel. As one extends the range of the UWB system, the delay spread increases, and the ISI gets worse.

There are two basic ways to address delay spread–induced ISI in UWB systems: using sophisticated equalizers to mitigate the effects of ISI or increasing the guard time between UWB pulses to prevent ISI in the first place. SK enables the latter. The high information content of each symbol in SK permits the use of low PRF for a given data rate. This, in turn, increases the guard time, preventing ISI from occurring and resulting in simplified implementation architectures.

SK has two disadvantages. First, it uses a high-order symbol, where the integrity of the entire symbol must be maintained. This is factored into the performance projections presented in this chapter. Second, the SK receiver must be capable of receiving each sub-band to demodulate the symbol. In general, this implies multiple RF paths.

7.2.3 Scalability to Higher Data Rates

One advantage of SK highlighted in the last section is its ability to scale to very high data rates. This is important within the context of the growing demand for high bandwidth connectivity. Table 7.1 shows the data rate for SK/QPSK systems at two PRFs. High data rates in excess of 1 Gbps can be obtained.

Spectral Keying™: A Novel Modulation Scheme for UWB Systems

Number of Frequency Bands	Data Rate for SK/QPSK at 12 MHz PRF (Mbps)	Data Rate for SK/QPSK at 24 MHz PRF (Mbps)
2	60	120
3	103	206
4	151	302
5	203	406
6	258	516
7	316	632
8	376	752
9	438	876
10	502	1,004

Table 7.1 Projected data rates for SK/QPSK Systems that use two to ten frequency bands.

7.2.4 SK Capacity Prediction

Capacity calculation is essential to evaluate any new modulation approach's potential to transmit information efficiently. The following analysis shows that a high information rate can be achieved with SK to justify its use in UWB systems.

The capacity of a channel is the maximum rate at which information can be transmitted reliably across the channel. This rate depends on the set of allowable symbols, the noise model of the channel, and the type of decoder used. This fact was first noticed by C. E. Shannon in his seminal paper [4], and the textbook by T. M. Cover and J. A. Thomas [5] gives a nice introduction to the subject. This section considers the capacity of SK modulation.

7.2.4.1 Input Alphabet

When discussing information theory and coding, the input alphabet is defined to be the set of symbols that may be transmitted. For example, M-ary frequency shift keying (M-FSK) allows any of M frequencies to be transmitted in any particular time slot. Therefore, the input alphabet has M symbols. In SK, however, the limitation on frequency reuse requires us to group T time slots together into one symbol.

Define an input symbol X to be an $M \times T$ matrix with $X_{ij} \in \{-1, 0, 1\}$, with B nonzero entries, and at most one nonzero entry in any row or column. A nonzero entry at position i, j represents a pulse of frequency i at time j, and the sign of the entry represents the phase of the burst. Consider our definition of SK, which is based on using M frequencies, T time slots, B frequency bursts, and the phase of the burst if $P = 1$. In this case, the size of the input alphabet is the product of the number of ways to choose the B frequencies from the M possible, the number of ways to choose the B time slots from the T possible, and the number of ways of ordering the B frequencies in time. Using the phase of the burst to transmit information also multiplies the number of symbols by 2^B. This means the number of symbols is given by

$$N(M,T,B,P) = \binom{M}{B}\binom{T}{B} B! 2^{BP}$$

Some systems also require that the first time slot always have a burst, which means the number of symbols is given by

$$\tilde{N}(M,T,B,P) = \binom{M}{B}\binom{T-1}{B-1} B! 2^{BP}$$

It is well known that the channel capacity, C_{SK}, can be upper bounded using the number of symbols, N, with $C_{SK} \leq \log_2 N$ bits per symbol.

7.2.4.2 Output Alphabet

Since SK consists of T time slots of M-FSK with an input restriction, the first step in demodulation is processed by T banks of M correlators. Let us denote an output symbol of size $M \times T$ as the matrix $\mathbf{Y}$, where Y_{ij} is the output of the correlator for the ith frequency during the jth time slot. Furthermore, picking the most likely input symbol given $\mathbf{Y}$ may require as many as N (i.e., the number of input symbols) correlators. Section 7.5 gives an efficient implementation by using the Viterbi algorithm on an appropriately defined trellis.

Consider this modulation on an additive white Gaussian noise (AWGN) channel with symbol energy per noise, $E_S/N_0 = 1/\sigma^2$. If coherent demodulation is used, then $\mathbf{Y}$ is defined by $Y_{ij} = X_{ij} + N_{ij}$, where the noise N_{ij} is a zero-mean Gaussian random variable with variance $\sigma^2/2$. If noncoherent demodulation is used, then $\mathbf{Y}$ is defined by $Y_{ij} = |X_{ij} + N_{ij}|^2$, where N_{ij} is a zero-mean circularly symmetrical complex Gaussian random variable with variance σ^2.

7.2.4.3 Estimating the Capacity

Using the basic tools of information theory, we can find the capacity of this modulation in AWGN. The symmetry of the modulation allows us to assume that all modulation symbols should be used equiprobably. This gives a capacity of

$$C = \log_2 N - H(X|Y)$$

where N is the number of symbols, and $H(X|Y) = E[\log_2 P_r(X|Y)]$ is the conditional entropy of Y given X. This expectation uses Monte Carlo methods by simulating the channel and decoder pair.

In this case, $P_r(X|Y)$ is usually computed via Bayes' rule (assuming $P_r(X) = 1/N$) with

$$P_r(X=x\,|\,Y=y) = \frac{P_r(Y=y\,|\,X=x)}{\sum_{x\in\chi} P_r(Y=y\,|\,X=x)}$$

where

$$P_r(Y=y\,|\,X=x) = \prod_{i=1}^{M}\prod_{j=1}^{T} P_r(Y_{ij}=y_{ij}\,|\,X_{ij}=x_{ij})$$

If coherent demodulation is used, then component distribution is Gaussian with mean x_{ij} and variance $\sigma^2/2$, which gives

$$P_r(Y_{ij}=y\,|\,X_{ij}=x) = \frac{1}{\sqrt{\pi\sigma^2}} e^{-(y-x)^2/\sigma^2}$$

If noncoherent demodulation is used, then component distribution is Ricean, which gives

$$P_r(Y_{ij}=y\,|\,X_{ij}=x) = \frac{y}{\sigma^2} e^{-(|x|^2+|y|^2)/2\sigma^2} I_0\left(\frac{|x|\,|y|}{\sigma^2}\right)$$

where I_0 is the Bessel function of the zeroth order.

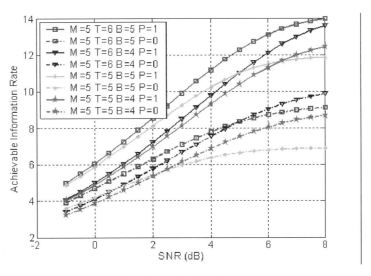

Figure 7.4 The capacity of coherently demodulated SK with M = 5.

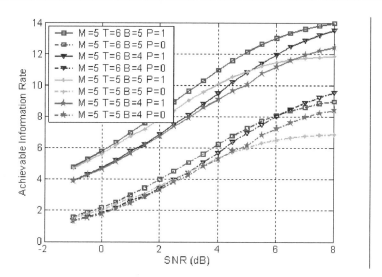

Figure 7.5 The capacity of noncoherently demodulated SK with M = 5.

Figures 7.4 and 7.5 show the results of example capacity calculations. The goal was to calculate the capacity of SK for the number of available bands $M = 5$, the number of available time slots $T \in \{M, M + 1\}$, the number of pulses per symbol $Q \in \{M - 1, M\}$, and polarity $P \in \{0, 1\}$. The results are broken into groups by M and demodulation type (i.e., coherent and noncoherent). Figure 7.4 gives the capacity of SK with $M = 5$ and coherent demodulation. Figure 7.5 gives the capacity of SK with $M = 5$ and noncoherent demodulation.

7.3 Performance Predictions

Another important metric for any modulation, in addition to an estimation of channel capacity, is the prediction of its performance under AWGN conditions. Performance bounds using soft decision decoding and AWGN channel are derived to predict SK performance. These bounds are based on representing SK as a nonlinear code over T time slots of M-FSK. The nonlinear code that defines SK is a symbol-wise geometrically uniform code, where the probability of symbol (or frequency) error is independent of the transmitted symbol frequency [6].

Let x_0 and x_1 be two constellation points, and let d be the squared Euclidean distance (SED) between them. The probability that the receiver chooses the symbol x_1 when x_0 was actually sent is given by

$$Q(\sqrt{d/2\sigma^2})$$

where σ^2 is the variance of the noise, and the Gaussian tail integral is given by

$$Q(x) = \frac{1}{\sqrt{2\pi}} \int_x^\infty e^{-t^2/2} dt$$

The probability of error for the optimal noncoherent detector is a function of the correlation coefficient ρ between x_0 and x_1 and can be written in terms of the Marcum Q function [7].

For the case where there are equal numbers of frequencies, time slots, and bursts ($M = T = B$) and no phase bits ($P = 0$), each symbol in the SK constellation can be associated with a permutation of the integers $\{1,\ldots, M\}$. For the SED between two constellation points that differ in f frequencies, the distance is $2f$, irrespective of the sent constellation point. It is noted that any two constellation points will differ by a minimum of two frequencies, giving a minimum SED of four. Let X_0 be the identity permutation (12345), and consider the number of constellation points that differ in f positions. This number, denoted by $A_f(M)$, is equal to the number that results from choosing f from M frequencies and deranging them. This means

$$A_f(M) = \binom{M}{f} d(f)$$

where $d(f)$ is the number of derangements of length f. Furthermore, the number of derangements can be computed using the recursion $d(n) = nd(n-1) + (-1)^n$, with the initial condition $d(1) = 0$. Using this weight enumerator, the word error rate bound is

$$P_w(M) \le \sum_f A_f(M) \; Q\!\left(\sqrt{\frac{fE_s}{N_0}}\right) \approx \frac{M(M-1)}{2} \; Q\!\left(\sqrt{\frac{2E_s}{N_0}}\right)$$

and the symbol error rate bound is then

$$P_s(M) \le \sum_f \frac{f}{M} A_f(M) \; Q\!\left(\sqrt{\frac{fE_s}{N_0}}\right) \approx (M-1) \; Q\!\left(\sqrt{\frac{2E_s}{N_0}}\right)$$

For the case with phase bit ($P = 1$) and $M = T = B$, the SED of two constellation points is two when frequency positions differ. If the frequencies match but phase differs, then the SED is four for that position. The number of constellation points that differ in f frequencies and p phase positions is then

$$A_{f,p}(M) = \binom{M}{f} d(f) 2^f \binom{M-f}{p}$$

Upper-bounding the frequency error rates (ignoring the phase) for the optimum detector, we get

$$P_w(M) \leq \sum_{f,p} \delta(f>0) A_{f,p}(M) \ Q\left(\sqrt{\frac{(f+2p)E_s}{N_0}}\right) \approx 2M(M-1) \ Q\left(\sqrt{\frac{2E_s}{N_0}}\right)$$

The corresponding symbol error rate bound is then

$$P_s(M) \leq \sum_{f,p} \frac{f}{M} A_{f,p}(M) \ Q\left(\sqrt{\frac{(f+2p)E_s}{N_0}}\right) \approx 4(M-1) \ Q\left(\sqrt{\frac{2E_s}{N_0}}\right)$$

For the case where we do not burst in all time slots ($B < T$), the SED is two and the word error rate is

$$P_w(T,B) \approx B(T-B) \ Q\left(\sqrt{\frac{E_s}{N_0}}\right)$$

Similarly, for the case where one or more frequencies are not sent ($B < M$), the SED is two and the word error rate is

$$P_w(M,B) \approx B(M-B) Q\left(\sqrt{\frac{E_s}{N_0}}\right)$$

Notice that for the last two cases, at high SNR, the word error rate is 3 dB worse than the $M = T = B$ case. Consequently, the best performance is achieved when $M = T = B$. Also, note the equation for this case is similar to antipodal modulation (e.g., BPSK). Simulations have confirmed the performance prediction in the equations above. Figure 7.6 gives a comparison of a MATLAB system simulation compared to the prediction in the equation above.

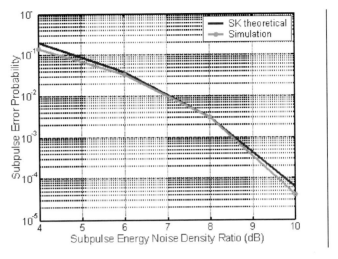

Figure 7.6 Comparison of BER performance of SK simulation and theoretical model.

7.4 SK Transmitter

The goal of the transmitter is twofold: (1) to encode the user data with redundancy in order to protect against noise and distortions acted upon the signal during transmission, and (2) to map the digital sequence into physical waveform. As mentioned earlier, SK modulation is considered a high-order modulation where a high number of bits can be mapped into one symbol. This section presents the symbol encoding and the transmit waveform of the SK modulation.

7.4.1 Transmitter Block Diagram

Figure 7.7 shows the transmit data path. The source data is encoded with a forward error correction (FEC) encoder of rate k/n for error protection and then scrambled by an interleaver for despreading bursty errors. In practice, channel impairment often impacts multiple received symbols and, as a result, produces a sequence of errors; when deinterleaved, the error sequence is spread over many bits and consequently improves the error correctability of the overall receiver. Following the interleaver, the bits-to-symbol block maps the digital bit sequence into SK symbols.

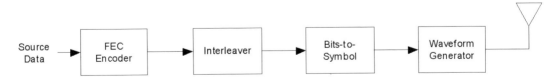

Figure 7.7 Transmit block diagram.

7.4.2 Mapping Bits to Symbols

The number of bits in a symbol in SK modulation is typically not a power of two since it is the $\log_2$ of the number of sub-bands. Consequently, various methods of mapping bits to symbols and vice versa must be considered. For example, a system with $M = 5$, $T = 5$, $B = 5$, and $P = 0$ has a symbol set of 120 members, generating $\log_2 120 = 6.907$ bits per symbol. If conversion efficiency is the goal, large blocks of bits should be mapped to large blocks of SK symbols.

Consider the following two examples: Mapper 1 takes 13 bits and generates two symbols, which gives an efficiency of 94.1%; Mapper 2 takes 27 bits and generates four symbols, which gives an efficiency of 97.7%. Mapper 2 clearly has better efficiency, but suffers from error propagation, larger implementation complexity, and larger decoding latency. As such, the general rule of thumb is to map a bit into no more than two symbols, which is depicted in Mapper 1.

A simple, but efficient, approach to mapping is to consider any SK symbol as a sequence of independent choices and map them to bit patterns. For the above example, the first received frequency can be mapped to 2 bits based on the choice between the 5 frequencies (in this case, one of the frequencies is not allowed to be sent first); the second received frequency can be mapped to 2 bits based on the choice between the four remaining frequencies; the third to 1.5 bits (by mapping a pair of third received frequency from 2 symbols to 3 bits), and the fourth to 1 bit. This gives 6.5 bits per symbol, which is 94 percent efficient. This mapping seems like a nice solution because it is fairly simple, reasonably efficient, and amenable to independent demodulation and decoding.

7.4.3 Waveform Generator

Figure 7.8 shows the block diagram for one possible implementation of an SK waveform generator. It relies on generating a carrier waveform centered on the sub-band.

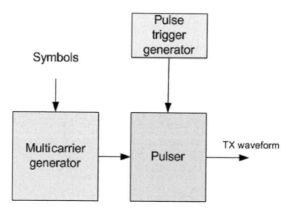

Figure 7.8 SK waveform generator.

The sub-band is selected based on the symbol being transmitted. The pulser generates a transmitted waveform.

Mathematically, the transmit waveform can be denoted as the sum of multiple pulsed waveforms

$$s(t) = \sum_{i=1}^{N} s_i(t - (i-1)T_s)$$

where N is the number of symbols, $s_i(t)$ is the waveform for the ith symbol, T_s is the symbol period, and $T_s = 1/\text{PRF}$. $s_i(t)$ can be further broken down as the linear combination of pulsed sinusoids

$$s_i(t) = \mathbf{m}'(t)\mathbf{X}_i\mathbf{p}(t)$$

where

$$\mathbf{m}(t) = \begin{bmatrix} \cos(\omega_1 t + \theta_1) \\ \vdots \\ \cos(\omega_M t + \theta_M) \end{bmatrix} \text{ and } \mathbf{p}(t) = \begin{bmatrix} p(t) \\ \vdots \\ p(t - (T-1) \cdot T_{sp}) \end{bmatrix}$$

$\mathbf{m}$ is the $M \times 1$ modulation vector; $\mathbf{X}_i$ is the ith symbol with dimension $M \times T$ as shown in Figure 7.3, and $\mathbf{p}$ is the $T \times 1$ pulse-shaping vector. T_{sp} is the sub-band pulse spacing. Here, $p(t)$ is selected to minimize adjacent band interference and to satisfy the FCC spectral mask requirement. Possible pulse shapes include Gaussian, rectangular, and raised cosine. Figure 7.9 depicts a possible transmit pulse waveform in both the time and frequency domains.

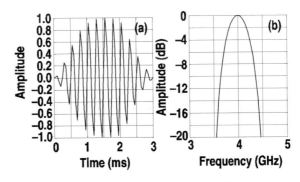

Figure 7.9 Gaussian transmit pulse shaping of a 4 GHz pulse: (a) time-domain and (b) frequency-domain representations.

7.5 Optimum Receiver for SK

7.5.1 Optimal Detection Theory

In the field of digital communications, the detector is the part of the system that converts noisy observations into hard or soft symbol decisions. In general, it is desirable to use a detector that optimizes some important parameters in the system. The complexity of optimal detection can be very large, which forces some systems to use suboptimal detection. An efficient method for optimal detection allows near-optimal performance at a reasonable complexity.

The optimal hard decision decoder minimizes the symbol error probability and is known as a maximum likelihood (ML) detector. One straightforward implementation of the ML detector involves making a list of all possible transmitted symbols, along with the probability of the received waveform given that symbol was transmitted. Picking the symbol in this list with the largest probability of being transmitted has the effect of minimizing the probability of symbol error (when all symbols are transmitted with equal probability). The complexity of this method grows linearly with the number of symbols.

The optimal soft decision detector provides sufficient statistics to the next stage of decoding and is known as an a posteriori probability (APP) detector. It takes this name because it computes, for each symbol, the probability that the symbol was transmitted given the received waveform (i.e., the APP). A straightforward implementation of the APP detector involves making a list of all possible transmitted symbols along with the probability that the symbol was transmitted given the received waveform. In many cases, each modulation symbol has an input label that consists of several subsymbols. So, the soft decision output of the detector is equal to the probability of each subsymbol, given the received waveform. This output can be formed by summing the probabilities of all symbols with a particular subsymbol value. The complexity of this method also grows linearly with the number of symbols.

7.5.2 Efficient ML Detection

Our efficient method of ML detection for SK is based on a graphical representation of the modulation known as a trellis diagram. Figure 7.10 shows the trellis diagram for an SK with $M = T = B = 3$, where M, T, and B were defined in Section 7.2.4.1. Using the standard assumption that the true waveform is corrupted by AWGN, we find that the ML detection problem is equivalent to minimum-distance detection, meaning that the symbol most likely to have been transmitted is the one that is closest in Euclidian Distance (ED) to the received waveform.

This section describes the efficient ML detector. Assume a simplified discrete-time model based on a bank of M matched filters (i.e., one for each frequency) that are coherently integrated. Let an SK symbol $u \in U$ be a vector of length T consisting of the frequencies transmitted at each symbol time. This means that $0 \le u_i \le M$ is the index of the frequency transmitted at time t, where $u_i = 0$ implies that nothing was transmitted at time t. Let the noiseless integrator output $x \in X$ be the $M \times T$ matrix given by

$$x_{i,t} = \begin{cases} 1 & \text{if } u_t = i \\ 0 & \text{otherwise} \end{cases}$$

This matrix corresponds to the integrated outputs of the M matched filters for T consecutive time slots. The matrix of integrator outputs $y \in Y$ is therefore given by $y_{i,t} = x_{i,t} + n_{i,t}$, where the noise $n_{i,t}$ is an independent zero-mean Gaussian random variable with variance σ^2. The minimum-distance symbol is defined to be

$$\hat{x} = \arg\min_{x \in X} \left[\sum_{t=1}^{T} \sum_{i=1}^{M} (y_{i,t} - x_{i,t})^2 \right]$$

Using BPSK on top of this system adds one bit of phase information per burst. This implies that $x_{i,t}$ can be ± 1. In this case, the minimum-distance detector starts by picking the stronger phase of each integrator output and then does minimum-distance detection. Stripping the phase from the integrator outputs is as simple as taking the absolute value, and this gives

$$\hat{x} = \arg\min_{x \in X} \left[\sum_{t=1}^{T} \sum_{i=1}^{M} (|y_{i,t}| - x_{i,t})^2 \right]$$

Using two bits of phase information implies that $x_{i,t}$ becomes a complex quantity with possible values of 1, –1, i, and –i. The same equation as above may be used, where the absolute value is replaced by the amplitude for the complex quantity.

A trellis diagram can be constructed for SK by keeping track of which frequencies can still be transmitted without violating the modulation constraint. In Figure 7.10, the state on the left denotes the initial state, {123}, because any of the three frequencies can be transmitted during the first burst. The state transition caused by transmitting the first burst corresponds to removing that frequency from the set. For example, if frequency 2 is transmitted, then the new state is {13}. Continuing this process until the empty set is reached gives the entire state diagram.

Minimum-distance detection of SK can be done very efficiently using this trellis representation. It starts by adding to each trellis edge the distance increment associated with choosing that trellis edge. Consider all the edges corresponding to time step t. If no phase information is transmitted, the distance increment associated with choosing any edge with output label i at time t is given by

$$d_{i,t}^{NP} = (y_{i,t} - 1)^2 - y_{i,t}^2 + \sum_{j=1}^{M} y_{j,t}^2$$

If phase information is used, then this is given by

$$d_{i,t}^{P} = (|y_{i,t}| - 1)^2 - y_{i,t}^2 + \sum_{j=1}^{M} y_{j,t}^2$$

Since the relative order of the distances is unchanged by subtracting the same constant from all edges at time t, the final sum in these expressions can be dropped. The relative order is also unchanged by subtracting the same constant from all trellis edges or scaling all trellis edges by a positive factor. This allows us to subtract one from all distance increments and divide them by two. The result of these modifications is that the weight of a trellis edge with output label i at time t is $w_{i,t} = -y_{i,t}$ (with no phase information) or $w_{i,t} = -|y_{i,t}|$ (with BPSK phase information).

Once the edge weights are defined, the minimum-distance detection algorithm simply finds the path from the leftmost state to the right state that minimizes the sum of all edge weights. This algorithm begins by initializing the leftmost state to zero. It then proceeds to extend the minimum-distance path one step to the right by choosing, for each state, the edge that minimizes the sum of the previous state metric and the edge weight. At each step, the algorithm keeps track of the input symbols associated with each best path.

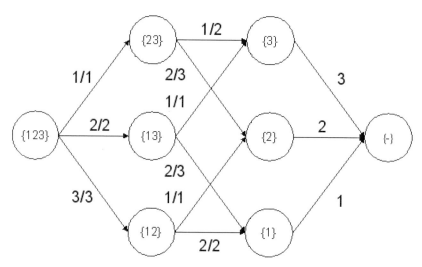

Figure 7.10 Trellis diagram for an M = T = B = 3 SK system.

7.5.3 Example of ML Implementation

Referring to Figure 7.10, each state (i.e., circle) is labeled with the frequencies that have not been transmitted yet. Each edge is labeled by i/j, where i is the input label, and j is the index of the transmitted frequency associated with that state transition. All valid SK symbols can be enumerated by concatenating the edge labels of left-to-right paths through this graph. The algorithm starts by initializing the state metrics $S(\bullet)$ and the path memories $M(\bullet)$. Initializing the state metrics gives $S(\{23\}) = -|y_{1,1}|$, $S(\{13\}) = -|y_{2,1}|$, and $S(\{12\}) = -|y_{3,1}|$. This initialization corresponds to taking the first trellis step because each state in the second trellis step has only one incoming edge. Initializing the path metrics gives $M(\{23\}) = \{1\}$, $M(\{13\}) = \{2\}$, and $M(\{12\}) = \{3\}$. This initialization allows us to keep track of the input symbols associated with the minimum-distance path. Looking backwards in the graph from state $\{3\}$, the optimal path to state $\{3\}$ can be extended either from state $\{23\}$ or $\{13\}$. Choosing the best path to state $\{3\}$ corresponds to picking the smaller of the two values $S(\{23\}) - |y_{2,2}|$ and $S(\{13\}) - |y_{1,2}|$. The path memory $M(\{3\})$ is computed by concatenating the input label of the winning edge to the path memory of the previous state. This step is known as the Add-Compare-Select operation, and it simply picks the path with the smaller metric and remembers which path was chosen. This process is continued until the terminal state on the left-hand side (LHS) is reached. The path memory associated with the minimum-distance path gives the ML codeword.

The minimum-distance decoding of an $M = T = B = 3$ SK system with BPSK on each frequency is described as an example. Suppose the integrator outputs are given by the following 3×3 matrix:

$$y = \begin{bmatrix} 1.0 & 2.0 & 2.0 \\ -1.5 & -2.0 & 1.0 \\ 3.0 & 3.0 & -1.0 \end{bmatrix}$$

Trying all six possible symbols shows that the symbol [3 2 1] with phases [+ − +] is the minimum-distance symbol. Trellis decoding begins by initializing the state metric $S(\{123\}) = 0$ and the path memory $M(\{123\}) = \emptyset$. Since there is only one edge to each of the next states, the partial distance for each state is given by $S(\{23\}) = -1$, $S(\{13\}) = -1.5$, and $S(\{12\}) = -3$. The path memories are also easily computed to be $M(\{23\}) = 1$, $M(\{13\}) = 2$, and $M(\{12\}) = 3$. The minimum-distance paths can now be extended to the next time slot. First, extend the minimum-distance path to state $\{3\}$ by writing $S(\{3\}) = \min(-3,-3.5) = -3.5$ and taking the input label from the winning path to get $M(\{3\}) = \{21\}$. Next, extend the minimum-distance path to state $\{2\}$ by writing $S(\{2\}) = \min(-4,-5) = -5$, which gives $M(\{2\}) = \{31\}$. Finally, extend the optimal path to state $\{1\}$ with $S(\{1\}) = \min(-5,-4.5) = -5$, where the winning input label gives $M(\{1\}) = \{32\}$. Finding the best path to the rightmost state gives $S(\{\ \}) = -7$ and gives our minimum-distance symbol as $M(\{\ \}) = \{321\}$.

7.5.4 Receiver Architecture for Multipath Channels

Section 7.5.3 derived an ML solution for SK in the AWGN channel. As described in other chapters, UWB systems will operate in multipath channels where the severity of the multipath varies depending on the application and the environment. While UWB systems will not suffer the deep fading faced by narrowband systems, ISI induced by multipath delay may still cause incorrect symbol decisions. Also, the transmitted signal energy will be dispersed into a large number of rays, causing a low percentage of that energy to be collected in the receiver.

The first problem has been classically solved using equalizers [7]. For good performance, equalizers are implemented in the digital domain. Considering that UWB systems are sampled at rates higher than 500 MHz to satisfy Nyquist sampling, these equalizers will be running at very high rates, increasing system complexity and power consumption. In SK, the need for equalizers is mitigated by providing a long-enough intersymbol guard period to allow the multipath of the previous symbol to decay before the arrival of the next symbol.

The second problem is fairly unique to UWB systems and is resolved with a diversity receiver implemented as a RAKE receiver. A diversity receiver relies on multiple receiver arms that receive statistically independent versions of the signal and is combined with a maximal ratio or equal ratio combiner.

In UWB systems, diversity is easily achieved since the signal bandwidth is typically much higher than the coherence bandwidth of the channel. Looking in the time domain, the UWB pulses are unlikely to fade completely due to multipath. So, the diversity receiver simply collects the multiple time-delayed versions of the received pulses. Consequently, in SK, implementing a RAKE receiver will require minimal extra hardware.

Figure 7.11 shows an SK receiver with a RAKE combiner. The SK receiver requires a separate arm for each sub-band. A matched filter is applied to the pulse in each sub-band. Optionally, a RAKE combiner coherently adds multiple rays that increase the collected signal energy.

7.5.5 Acquisition and Channel Sounding

Acquisition of the SK signal is straightforward. One implementation uses an acquisition preamble made up of a predefined sequence of symbols. The number of symbols in the preamble will depend on whether the symbols are combined in a coherent or noncoherent fashion and whether soft or hard decisions on the symbols are used. A symbol sequence can also be used as an end-of-preamble delimiter.

Acquisition can also be used to estimate the channel parameters useful for the receive function. As before, the channel parameters will depend on the type of demodulator used. A noncoherent receiver will only require estimation of the rela-

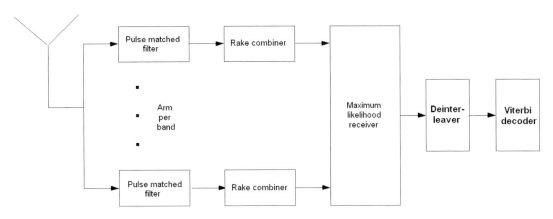

Figure 7.11 Receiver block diagram.

tive channel delays between the multiple sub-bands. A coherent receiver will require additional parameters, such as a phase reference for the continuous receive function and an estimate of the frequency error between the transmitter and receiver.

7.5.6 Implementation Scalability

One of the major advantages of SK is the scalability of its receivers from a very simple implementation with acceptable performance to a near-optimum solution with acceptable complexity. For example, a simplified solution may use a noncoherent receiver by realizing the pulse-matched filter in Figure 7.11 with the combination of a bandpass filter and a square law or envelope detector. The ML receiver may be replaced with a simple threshold detector and so on.

7.6 Conclusion

This chapter has presented the novel modulation approach of SK. It embodies the multiband approach to UWB, which has been championed by General Atomics. This approach has already been demonstrated in working RF and baseband chips targeted for wireless video and wireless cable replacement markets.

Acknowledgement is given here to Henry Pfister, who helped with the analysis of the SK coding performance.

References

1. Askar, Naiel K., S. C. Lin, H. D. Pfister, G. E. Rogerson, and D. S. Furuno. "Spectral Keying™: A Novel Modulation Scheme for UWB Systems." UWBST, Ruston, Virginia (November 2003).
2. Askar, Naiel. General Atomics CFP presentation, doc: IEEE 802.15-03/105r1, March 2003.

3. Razzell, Charles. Philips CFP presentation. doc: IEEE 802.15-03/125.
4. Shannon, C. E. "A Mathematical Theory of Communication." *Bell System Technical Journal* 27 (July/October 1948): 379–423, 623–656.
5. Cover, T. M., and J. A. Thomas. *Elements of Information Theory.* New York: Wiley, 1991.
6. Forney, G. D., Jr. "Geometrically Uniform Codes." *IEEE Trans. Inform. Theory* 37, no. 5 (September 1991): 1241–1260.
7. Proakis, J. G. *Digital Communications, 4th ed.* New York: McGraw-Hill, 2001.

8

Multiband OFDM

by Jaiganesh Balakrishnan and Anuj Batra

Over the last 10 years, there has been some amount of interest in UWB communication systems, but the real explosion in the research and design of these systems did not occur until the 2002 landmark ruling by the Federal Communications Commission (FCC) [1–3]. In this ruling, the FCC allocates 7,500 MHz (from 3.1 to 10.6 GHz) of unlicensed spectrum for use by UWB communication devices. Much of the increased interest in these devices is due to the availability of allocated spectrum, as well as to the ability of UWB devices to deliver data rates that can scale from 53.3 Mbps at a distance of 15m up to 480 Mbps at a distance of 2m in realistic multipath environments, all while being a low-cost and low-power-consumption solution. It is expected that these UWB devices will satisfy the consumer's insatiable appetite for data rates as well as enable new consumer market segments.

In addition to allocating spectrum, the FCC's ruling also specifies a minimum instantaneous bandwidth (at the 10 dB points) of 500 MHz for the UWB signal. This requirement has revolutionized the design of many UWB communication systems. Instead of using the entire band to transmit information, the spectrum can now be subdivided into several bands with bandwidth of approximately 500 MHz, and the information symbols can be time-interleaved across multiple bands. There are two main advantages to time-interleaving the symbols: (1) the effective total average transmit power is equal to average power transmitted per sub-band times the number of bands, and (2) the instantaneous processing bandwidth is smaller (approximately 500 MHz) than using the entire bandwidth, which leads to lower power consumption, lower cost, and improved spectral flexibility for worldwide compliance and coexistence. Other advantages of this approach include using lower-rate analog-to-digital converters (ADCs) and simplifying the digital complexity. UWB systems built using this approach are often referred to as multiband systems.

Ultimately, the FCC ruling has helped to create many new standardization efforts, such as Ecma TC32-TG20 and IEEE 802.15.3a, which have focused on developing high-speed wireless communication systems for personal area networks. In addition to standardization efforts, the ruling has led to new opportunities for innovation and technical advancement.

The remainder of this chapter is organized as follows. Section 8.1 provides a brief description of previous UWB approaches and their main advantages and disadvantages. This section serves as motivation for the design of a new approach to high-speed UWB communications, often referred to as multiband orthogonal frequency-division modulation (MB-OFDM). Section 8.2 investigates the optimal operating bandwidth for a UWB system. The design trade-offs leading to the modulation choice for the MB-OFDM system are described in Section 8.3. The various design trade-offs and optimal design parameters for the MB-OFDM system are discussed in Section 8.4, as are the complexity and power consumption estimates of the MB-OFDM system. Performance results based on detailed link level simulations are provided.

8.1 Overview of Previous UWB Approaches

Traditional designs of UWB communication systems involve using very narrow time-domain pulses, on the order of 500 ps to 2 ns, to generate signals that occupy an extremely wide spectrum [4]. The information is then transmitted by modulating the pulses in either time, for instance, using pulse-position modulation (PPM), or in phase, for instance, using binary phase-shift keying (BPSK). In addition, either spreading or time hopping is added to the transmitted signal in order to prevent spectral lines in the transmitted power spectral density (PSD). The main advantage of this approach is that the transmitted signal can easily be generated in the analog domain using analog circuits. The main disadvantage of this approach is that the analog circuits and mixed-signal circuits, such as the ADCs, in the receiver are difficult to design when the bandwidth of the signal is extremely large. In addition, these circuits will likely consume a large amount of current in order to process the signal at a high rate and maintain a sufficiently low noise figure. The other main challenge of this approach is that it requires a large number of RAKE fingers in order to capture sufficient energy in a dense multipath environment, thereby significantly increasing the complexity of the digital baseband for the physical layer.

Another approach, often referred to as pulsed multiband, eliminates many of the disadvantages associated with UWB systems that use very narrow time-domain pulses; specifically, it eliminates the need to process a very large bandwidth. The pulsed multiband approach divides the spectrum into several bands with a bandwidth of approximately 500 MHz. The information is then modulated using PPM or BPSK and transmitted on each band using narrow time-domain pulses, on the order of 2 to 4 ns [5]. By interleaving the symbols across the bands, the pulsed multiband can maintain the same transmit power as if it were using the entire bandwidth. The time between symbols—correspondingly, the dwell time in each center frequency—is typically on the order of 4 to 8 ns. The main advantage of this approach is that the information can now be processed over a much smaller bandwidth, reducing the complexity of the design and the power consumption, lowering the cost, and improving spectral flexibility and worldwide compliance.

The primary disadvantage of the pulsed multiband approach is the difficulty of collecting significant multipath energy using a single RF chain. Multipath energy collection is an important issue because it fundamentally determines the range of a communication system. The amount of multipath energy that can be collected by a pulsed multiband system with a single RF chain is limited to the dwell time for each band, on the order of 4 to 8 ns, which is a small fraction of the total delay spread for most UWB channels. Typical non-line-of-sight (NLOS) channels have total delay spreads that range from 40 to 70 ns. If the system cannot capture significant multipath energy, then it will not be able to achieve the range promised by UWB communication systems. The multipath energy collection can be improved by designing a pulsed multiband system with several RF chains; however, this comes at the expense of increased cost and power consumption.

Another disadvantage of the pulsed multiband system is that when it employs a small number of RAKE fingers, its performance is very sensitive to group delay variations introduced by the analog front-end components. In addition, the pulsed multiband system places very stringent frequency-switching time requirements, on the order of 100 ps, at both the transmitter and receiver.

Many of the disadvantages of pulsed multiband systems can be overcome by using a symbol that has a much longer duration and choosing a modulation technique that can efficiently capture multipath energy. By combining orthogonal frequency-division modulation (OFDM) with multibanding, a new approach can be developed that inherits all the strengths of the pulsed multiband approach while addressing the issue of multipath energy capture. Often referred to as MB-OFDM, this approach uses the OFDM symbols to convey the information on each of the bands and interleaves them across all the bands in order to achieve the same power as an approach that uses the same total bandwidth instantaneously. The MB-OFDM approach has several nice properties, including the ability to capture multipath energy efficiently with a single RF chain, insensitivity to group delay variations, the ability to deal with narrowband interferers at the receiver without having to sacrifice either bands or data rate, and spectral flexibility. In addition, this approach offers relaxed frequency-switching time requirements, on the order of 9 ns, as compared to the pulsed multiband approach. The only drawback to this type of approach is that the transmitter is slightly more complex because it requires an inverse discrete Fourier transform (DFT), and the peak-to-average ratio may be slightly higher than that of the pulse-based multiband approaches.

8.2 Optimal Operating Bandwidth

An important design parameter for UWB systems is the choice of initial operating bandwidth. The FCC ruling limits the transmitted radiated PSD to −41.3 dBm/MHz. Therefore, the total radiated power is directly related to the operating bandwidth. Clearly, using more bandwidth yields a higher transmit power. At the same time, increasing the operating bandwidth also complicates the design of the low noise amplifier (LNA) and mixer, as well as requires higher-speed digital-to-analog converters (DACs) and ADCs and correspondingly higher-speed digital logic.

The optimal operating bandwidth can be determined via a simple analysis, which examines the two primary entries in the link budget table that are affected by operating bandwidth. Under the ideal assumption that the transmitted PSD is flat, the received signal power P_{RX} can be expressed as follows:

$$P_{RX}(f_U - f_L) = -41.25 + 10\log_{10}(f_U - f_L) - P_L(f_g, d) \text{ (dBm)}$$

where the lower frequency, f_L, is defined as 3.1 GHz per the FCC ruling, and the upper frequency, f_U, is varied from 4.8 to 10.6 GHz. $P_L(f_g, d)$ represents the path loss between two devices and is defined as follows:

$$P_L(f_g, d) = 20\log_{10}\left[\frac{4\pi f_g d}{c}\right]$$

where f_g is the geometric mean of f_L and f_U, d is the distance in meters, and c is the speed of light. Although the path-loss equation assumes free space propagation, the conclusion is independent of the exact path-loss model.

Figure 8.1 shows the received signal power at a distance of 10m as a function of upper frequency f_U. This figure clearly shows that the received signal power increases by 2.0 dB at f_U = 7.0 GHz when compared to f_U = 4.8 GHz. Similarly, the received signal power increases by 3.0 dB at f_U = 10.6 GHz when compared to f_U = 4.8 GHz. On the other hand, the noise figure for the LNA will increase by at least 1.0 dB at f_U = 7.0 GHz when compared to f_U = 4.8 GHz. Similarly, the noise figure for the LNA will increase by at least 2.0 dB at f_U = 10.6 GHz when compared to f_U = 4.8 GHz. In either case, the overall link margin increases by at most 1.0 dB when the upper frequency of the operating bandwidth exceeds 4.8 GHz. However, this increase in link budget comes at the expense of higher complexity, longer design cycles, and increased power consumption at the receiver.

Limiting the upper frequency of the operating bandwidth to 4.8 GHz has several advantages, including shortening the time to market, simplifying the design of the RF and analog front-end circuits, specifically LNAs and mixers, and making the technology more friendly towards implementation in CMOS technology. Another important design consideration in selecting the operating bandwidth is that interferers may potentially lie within the bandwidth of interest. For example, in the United States, the U-NII band occupies the bandwidth from 5.15 to 5.85 GHz, while in Japan, the U-NII band occupies the bandwidth from 4.9 to 5.1 GHz. By limiting the upper frequency to 4.8 GHz, these interferers can be avoided, and it is also possible to simplify the design of the system by removing the need to implement complicated notch filters to suppress the interference.

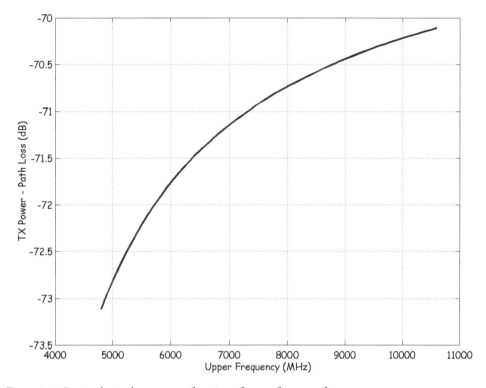

Figure 8.1 Received signal power as a function of upper frequency f_U.

Limiting the initial operating bandwidth to the first 1,700 MHz does not imply that the entire 7,500 MHz will not eventually be utilized. As RF technology improves, it will become more efficient to use the upper frequencies in the UWB range. If defined with forethought and proper planning, the UWB systems can accommodate an effective migration path to the upper end of the spectrum when market conditions dictate such a move.

8.3 Modulation Choice

The selection of modulation scheme for high-speed UWB communications is typically application dependent and driven by a number of parameters, including the key criteria of performance, complexity, and system flexibility. The performance of the UWB system is determined by its robustness to multipath channel environments and its ability to handle narrowband interferers and other UWB interferers. The ability to sculpt the transmit spectrum is also an important requirement as the UWB regulations have not been finalized in many countries.

Given an operating bandwidth from 3.1 to 4.8 GHz, there are several ways to design a high-speed UWB communication system. One method is to transmit the information using spread-spectrum or direct-sequence (DS) techniques [6] over the entire 1,700 MHz (see Chapter 6). The main advantage of this approach is that spread-spectrum techniques are well understood and have been proven in other commercial technologies (e.g., wideband CDMA). Of course, transmitting a signal over the entire 1,700 MHz does complicate the design of the RF and analog circuits, as well as of high-speed ADCs. In addition, the complexity of the digital baseband section of the physical layer can be quite large because 16 RAKE fingers are typically needed to capture sufficient multipath energy to meet the 10m range requirements that a data rate of 110 Mbps requires.

Another approach to designing a high-speed UWB communication system is to divide the operating bandwidth from 3.1 to 4.8 GHz into three bands, where each band has a bandwidth of 528 MHz. The information in each band is transmitted using an OFDM modulation scheme with a 128-point FFT. A cyclic prefix (CP) of 60.6 ns is preappended to the OFDM symbol in order to provide multipath robustness. The resulting OFDM symbols are then interleaved across the three bands to ensure that the radiated transmitted power is the same as that of a CDMA system.

In this section, the suitability of OFDM as the modulation choice is demonstrated by the fact that it satisfies the key criteria better than a single-carrier DS-UWB approach. For the sake of brevity, the following analysis is restricted to the single-carrier DS-UWB approach and an MB-OFDM system.

The remainder of this section analyzes the computational complexity and multipath energy collection capabilities of a single-carrier DS-UWB system operating at a chip rate of 1,368 MHz with a 16-finger RAKE and of a multicarrier system with a 128-point FFT, 60.6 ns CP, and three-band system, where each band has a bandwidth of 528 MHz.

8.3.1 Performance in Multipath Channels

The performance and robustness of a wireless communication system are often determined by the amount of multipath energy that can be collected at the receiver. UWB channels are highly dispersive and therefore pose significant challenges in the design of high-speed wireless communication systems [7–10]. Typical NLOS channel environments at distances of 4m to 10m have an RMS delay spreads of 14 ns, while the worst-case channel environments have an RMS delay spread of 25 ns [11].

For the single-carrier system, the most common way to collect multipath energy is to use a RAKE receiver. The performance of a single-carrier system in a highly dispersive UWB channel is limited by two effects: a large number of RAKE fingers is needed to capture multipath energy sufficiently and the time-dispersive nature of the channel causes intersymbol interference (ISI), resulting in performance degradation [12]. The effect of ISI can be mitigated by the use of an equalizer, but this performance gain comes at the cost of additional computational complexity.

Let the equivalent discrete-time baseband received sequence $r(n)$ be written as follows:

$$r(n) = \sum_{k=0}^{L-1} h(k)s(n-k) + w(n)$$

where $s(n)$ is the transmitted sequence, $h(k)$ is the channel impulse response of length L, and $w(n)$ is the noise sequence. Let $y(n)$ represent the output of the RAKE receiver with a span of L coefficients, of which only M of the coefficients, or "fingers," are nonzero. Let $d(k)$ represent the delays of M nonzero coefficients of the L-tap RAKE receiver response $f(n)$. Thus, the output of the RAKE receiver can be written as follows:

$$y(n) = \sum_{k=0}^{M-1} f[\delta(k)] r[n - \delta(k)]$$

All the multipath energy can be captured if the receiver filter response is matched exactly to the channel impulse response. However, this implies that M would have to be equal to L; increasing the length of the RAKE receiver would result in a significant increase in the complexity of the digital baseband.

Figure 8.2 shows the loss in captured multipath energy and the signal-to-ISI ratio for the CDMA system as a function of the number of RAKE fingers. The results in this figure are shown for the 90th percentile channel realization corresponding to a 4–10m NLOS channel environment (CM3) and a data rate of 114 Mbps.

These curves were generated assuming that the multipath channel impulse response is perfectly known at the receiver and that the RAKE fingers are placed to capture the largest multipath coefficients within a span of about 40 ms. This figure shows that the CDMA system can only capture 56 percent of the available multipath energy, even with the optimal placement of the 16 RAKE fingers. This figure also shows that with 16 RAKE fingers, the ISI term is only 9 dB below the signal energy. Having such a large ISI term will definitely degrade the system performance unless an equalizer is also included in the receiver. Note that as the data rate doubles, the processing gain reduces by a factor of two; consequently, the ISI term increases by 3 dB. Therefore, for a data rate of 228 Mbps, the ISI term is approximately 6 dB below the signal energy for a system with 16 RAKE fingers. With only 6 dB between the signal level and the ISI level, there will be an insufficient signal-to-interference-and-noise ratio (SINR) to decode the information bits successfully unless an equalizer, which will dramatically increase the complexity of the digital baseband, is used to mitigate the ISI.

On the other hand, Figure 8.3 shows the loss in captured multipath energy and the ratio of inter-carrier-interference (ICI) to signal for the MB-OFDM system, with an instantaneous bandwidth of 528 MHz, as a function of the CP length. This figure shows that the MB-OFDM system captures approximately 95 percent of the

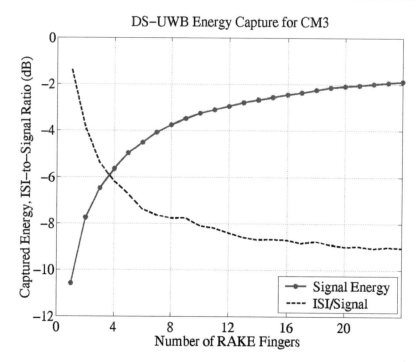

Figure 8.2 90th percentile multipath energy capture of a DS-UWB system for a 4–10m NLOS channel environment.

multipath channel energy with a CP length of 60.6 ns for the 90th percentile channel realization corresponding to a CM3 channel environment and a data rate of 106.7 Mbps.

8.3.2 Receiver Complexity

The complexity of the receiver is a critical parameter in determining which modulation scheme to use. The complexity of the single-carrier system increases linearly with the number of RAKE fingers and the receiver sampling rate.

For the DS-UWB system, an M-finger RAKE receiver requires M complex multiply operations for every chip processed. For example, a 16-finger RAKE receiver implemented at a chip rate sampling of 1,368 MHz requires 21.9 complex multiply operations every nanosecond. This complexity analysis only analyzes the complexity of the RAKE receiver and does not include the complexity of a high-speed equalizer, which is often needed in single-carrier systems at data rates greater than 200 Mbps.

The complexity of the OFDM system varies logarithmically with the FFT size. For an N-point FFT, $(N/2) \times \log_2(N)$ complex multiply operations are required every OFDM symbol. Note that the OFDM symbol is typically longer than N-samples due to the presence of a CP. For the MB-OFDM system, the FFT requires 1.48

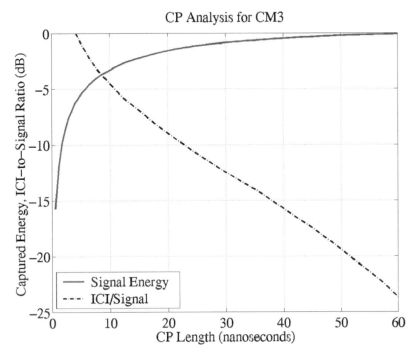

Figure 8.3 Captured multipath energy for an MB-OFDM system as a function of CP length for a 4–10m NLOS channel environment.

complex multiply operations every nanosecond.[1] In addition, the single-tap frequency-domain equalizer requires an additional 0.42 complex multiply operations every nanosecond, resulting in a total receiver complexity of 1.9 complex multiply operations per nanosecond. Note that the multicarrier system does not require any additional complexity to achieve the higher information data rates.

When the inverse of the sampling rate is significantly shorter than the total delay spread or, equivalently, for highly dispersive channels as is the case for most UWB channels, OFDM receivers are much more efficient at capturing multipath energy than an equivalent single-carrier system using the same total bandwidth.

In addition to capturing energy more efficiently, an OFDM system also possesses several other desirable properties: high spectral efficiency, inherent resilience to narrowband RF interference, and spectral flexibility, which is important because the regulations for UWB devices have not been finalized throughout the entire world. With an OFDM system, the transmitted spectrum can easily be shaped by nulling out tones or turning off bands in order to protect sensitive or critical bands.

---

[1] The FFT requires $64 \times \log_2(128)$ complex multiply operations every 160 samples at a sampling rate of 528 MHz.

8.4 MB-OFDM

This section provides an overview of the MB-OFDM system and discusses the corresponding design trade-offs when developing an OFDM-based UWB system. For the sake of simplicity, an MB-OFDM system using only the first three bands will be described, but the details about how to use the full set of 14 bands can be found in [13].

8.4.1 Architecture for an MB-OFDM System

The leading technology for use in high-speed wireless UWB communication systems is the MB-OFDM approach, which combines OFDM with multibanding. In this approach, the spectrum is divided into 14 sub-bands, each having a bandwidth of 528 MHz. Information is transmitted using time-domain OFDM symbols across three consecutive bands in a time-interleaved fashion, as illustrated in Figure 8.4 [14–19].

The MB-OFDM solution is very similar to many conventional wireless OFDM systems. However, many aspects of the MB-OFDM system were chosen to reduce the implementation complexity: (1) the constellation size is limited to quadrature phase-shift keying (QPSK), which reduces the internal precision of the digital logic (specifically the IFFT and FFT) and the required precision of the ADC and DAC; (2) a large subcarrier spacing is used, which relaxes the phase-noise requirements on the carrier synthesis circuitry and improves the system robustness to synchronization errors; (3) a zero-padded suffix (ZPS) is used instead of a CP to retain robustness against multipath, maximize transmit power, and allow sufficient time for the transmitter and receiver to switch carrier frequencies; and (4) a time-frequency kernel is used to determine the center frequency for the transmission of each OFDM symbol. An example of how the OFDM symbols are transmitted in an MB-OFDM is shown in Figure 8.4.

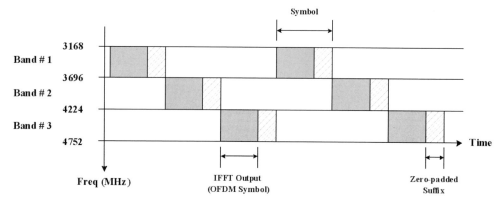

Figure 8.4 Example of time-frequency coding for an MB-OFDM system.

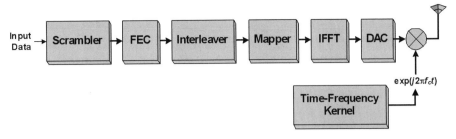

Figure 8.5 Example transmitter architecture for an MB-OFDM system.

Figure 8.4 shows one realization of a time-frequency code, where the first OFDM symbol is transmitted on band #1, the second OFDM symbol is transmitted on band #3, the third OFDM symbol is transmitted on band #2, the fourth OFDM symbol is transmitted on band #1, and so on. The time-frequency codes are used not only to provide frequency diversity in the system but also to enable multiple-access between piconets operating in the same vicinity.

An example of the architecture for an MB-OFDM transmitter [14] is shown in Figure 8.5. At the transmitter, the input bit stream is scrambled. A forward error correction (FEC) code is now applied to provide resilience against transmission errors. The encoded sequence is now interleaved and then mapped to frequency bins of an OFDM symbol. An IFFT is used to transform the frequency-domain information into a time-domain OFDM symbol. The OFDM symbols are then converted into continuous time-domain analog waveforms, up-converted to the appropriate center frequency, and then transmitted.

Figure 8.6 illustrates an example of the architecture for an MB-OFDM receiver. The received signal is amplified using an LNA and down-converted to the complex baseband using in-phase and quadrature mixers. The time-frequency kernel provides the sequence of sub-band center frequencies based on the appropriate time-frequency code. The complex baseband signal is low-pass filtered to reject out-of-band interferers. This is then sampled and quantized using a 528 MHz ADC to obtain the complex digital baseband signal.

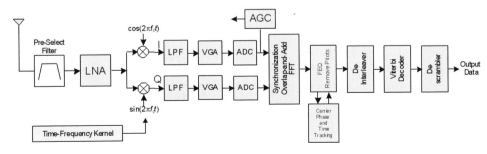

Figure 8.6 Example receiver architecture for an MB-OFDM system.

Processing of the baseband signal begins with the packet detection to determine if there is any packet-on-air. The automatic gain control (AGC) loop controls the settings of the variable gain amplifier (VGA) to maintain the best possible signal swing at the input of the ADC. As soon as the packet is detected, the receiver initializes the time-frequency kernel, starts searching for the sync frame sequence, and determines the optimal FFT placement window for each of the three sub-bands. The samples corresponding to the ZPS are added to the start of the OFDM symbol using the overlap-and-add module. An FFT operation is now performed to obtain the frequency-domain information, and the output of the FFT is equalized using a frequency-domain equalizer (FEQ). A phase correction is applied to the output of the FEQ to undo the effect of carrier and timing mismatch between the transmitter and receiver. The pilot tones in each OFDM symbol are used to drive the digital PLL. The output of the FEQ is demapped and deinterleaved before passing on the Viterbi decoder. The error-corrected bit sequence is now descrambled and passed on to the MAC.

8.4.2 Band Planning

Since the FCC specifies that a system must occupy a minimum of 500 MHz (10 dB) of bandwidth in order to be classified as a UWB system, the maximum number of bands that can be placed between 3.1 and 4.8 GHz is three. One possible option is to use a single band that spans the entire operating bandwidth. However, from an implementation point of view (lower power consumption for the channel select filter, lower speed for the ADCs, and lower speeds for the digital processing) and a multiple access point of view (larger selection of time-frequency codes), using a smaller-bandwidth band is much more desirable. Therefore, choosing a sub-band bandwidth as close to 500 MHz as possible is a better choice.

In the MB-OFDM system, a bandwidth of 528 MHz for each of the bands was chosen for two specific reasons. First, it allows sufficient guard band on the lower side of band #1 and the upper side of band #3 to simplify the design of a preselect filter, which is used to attenuate the out-of-band signals (including emissions from the GPS, GSM, PCS, ISM, and U-NII bands). Second, the center frequencies for the three-band system were chosen so as to simplify the design of the synthesizer and ensure that the system can switch between the center frequencies within a few nanoseconds. The band plan can also be extended to encompass the remaining portions of the UWB spectrum, as shown in Figure 8.7.

In this band plan, five nonoverlapping band groups are defined, where the first four band groups are formed using three consecutive bands, while the last band group is constructed using the last two consecutive bands. From an implementation perspective, Band Group #1 is mandatory, while the remaining band groups are optional. However, it is expected that MB-OFDM will eventually operate over all band groups and that applications may be separated by band group. For example, devices requiring larger range, such as DVDs to flat panel televisions, will prefer to

Multiband OFDM

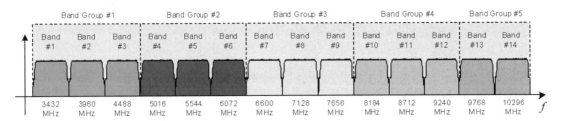

Figure 8.7 *Frequency-band plan for the MB-OFDM system.*

use band groups #1 and #2, while devices that can live with shorter ranges, such as digital still cameras to laptops, will naturally gravitate to band groups #3 and #4.

8.4.3 Synthesizer Architecture

Figure 8.8 shows an example of a synthesizer architecture that can rapidly switch between the three bands within band group #1. This architecture exploits the relationship between the center frequencies for the three-band system and the oscillator frequency, namely 4,224 MHz. The basic idea is that each of the center frequencies is generated using a single-sideband beat product of the oscillator frequency with another frequency derived from the oscillator. The other frequency may be obtained from the oscillator by using a combination of frequency dividers and single-sideband mixers. For example, the center frequency for band #1 is generated by mixing 4,224 MHz with 792 MHz to obtain a frequency of 3,432 MHz. The 792 MHz signal is generated by mixing 528 MHz (= 4,224/8) and 264 MHz (= 4,224/16). The center frequencies for bands #2 and #3 can be obtained by mixing 4,224 MHz with 264 MHz (= 4,224/16).

The advantage of this architecture is that all of the center frequencies can be generated from a single PLL. Since all of the center frequencies are available at all times, switching between the different sub-bands can be accomplished within a few nano-

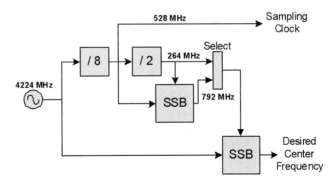

Figure 8.8 *An example synthesizer architecture that can switch between frequencies within a few nanoseconds.*

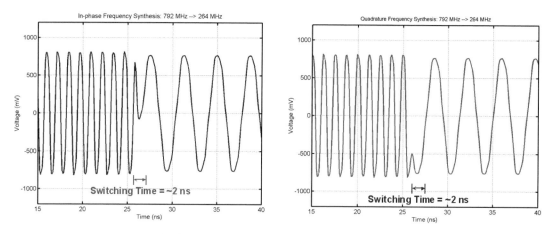

Figure 8.9 Circuit-level simulations of the frequency-switching architecture.

seconds. In fact, the exact time required for switching between frequencies is limited by the response time of the multiplexer.

Figure 8.9 illustrates the plot from a circuit-level simulation of the frequency-switching architecture shown in Figure 8.8. This plot clearly shows that the nominal switching time between frequencies is approximately 2 ns.

8.4.4 Number of Tones

Another important design parameter for an OFDM system is the size of the FFT. This block typically accounts for 25 percent of the complexity of the receiver digital baseband. In order to have a low-cost solution, it is important to keep the number of tones for the FFT as small as possible. For the initial design of the MB-OFDM system, FFT sizes of 64 points (~51,000 gates) and 128 points (~70,000 gates) were considered. Since the MB-OFDM is targeted toward portable and handheld devices, an FFT size of 256 points (~91,000 gates) was considered borderline too complex for a low-cost, low-complexity solution. There is a natural trade-off when selecting the number of tones for the FFT; a smaller number of tones results in a larger overhead due to the CP and reduces the range of the system, while a larger number of tones results in a lower overhead due to the CP but increases the complexity of the solution.

Figure 8.10 shows a performance comparison between an MB-OFDM system using a 64-point FFT and a prefix length of 41.7 ns, a system using a 64-point FFT and a prefix length 54.9 ns, and a system using a 128-point FFT and a prefix length of 60.6 ns for a data rate of 106.7 Mbps in a CM3 channel environment as a function of distance. A path-loss decay exponent of two (free space propagation) was assumed for simulations. The packet error rate (PER) curves were averaged over at least 500 packets, where each packet carried a payload of at least 1,000 octets. Note

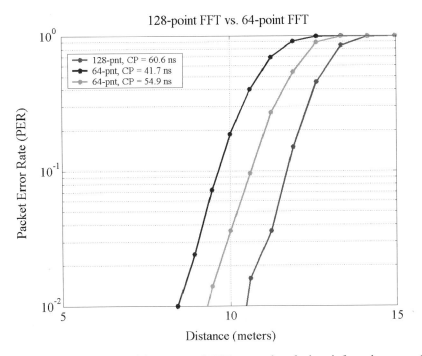

Figure 8.10 PER as a function of distance, and FFT size and prefix length for a data rate of 106.7 Mbps.

that the performance simulations incorporate losses due to front-end filtering, clipping at the DAC, ADC degradation (four bits for 110/200 Mbps and five bits for 480 Mbps), multipath, shadowing, packet acquisition, channel estimation, clock frequency mismatch (±20 ppm at the TX and RX), carrier offset recovery, carrier tracking, and so forth. The results shown in this figure correspond to the performance of the 90th percentile best channel realization (i.e., the PER curves for the worst 10 percent of channels are discarded).

This figure shows that a system using a 128-point FFT with a 60.6 ns prefix is approximately 0.9 dB better than a system using a 64-point FFT with a 54.9 ns prefix and approximately 1.7 dB better than a system using a 64-point FFT with a 41.7 ns prefix. Thus, the optimal size for the FFT for the MB-OFDM system is 128 points, which provides an excellent balance between performance and complexity.

8.4.5 Prefix Length

An OFDM system offers inherent robustness to multipath dispersion with a low-complexity receiver. This property is a result of the addition of a CP, which forces the linear convolution with the channel impulse response to resemble a circular convolution [20]. A circular convolution in the time domain is equivalent to a multiplication operation in the DFT domain. Hence, a single-tap frequency-domain equalizer is sufficient to undo the effect of the multipath channel on the transmitted signal.

The length of the CP determines the amount of multipath energy captured. Multipath energy not captured during the CP window results in ICI. Therefore, the CP length needs to be chosen to minimize the impact due to ICI and to maximize the collected multipath energy, while limiting the overhead due to the CP. From Figure 8.3, it is clear that a CP duration of 60.6 ns maximizes the capture of multipath energy and minimizes the impact of ICI for all channel environments. Note that the ICI-to-signal ratio shown in this figure is calculated at the input of the Viterbi decoder and, hence, incorporates the processing gain that is expected from a data rate of 106.7 Mbps.

While most conventional wireless OFDM systems use a CP to provide robustness against multipath, the same multipath robustness can be obtained by using a ZPS instead of the CP [21]. At the receiver, the only modification that is required to maintain the same multipath robustness the collection of additional samples corresponding to the length of the suffix and the use of an overlap-and-add method to obtain the circular convolutional property.

The main advantage of using a ZPS is that the transmit power can be maximized. When a CP is used, redundancy or structure is introduced into the transmitted signal, which leads to ripples in the average PSD. Since the maximum radiated transmit power for UWB systems is limited by the FCC, any ripples in the transmitted PSD will require some amount of reduction in the transmitted power. The amount of transmit power reduction required is equal to the peak-to-average ratio of the PSD. For the MB-OFDM system, this reduction in power could be as large as 1.5 dB, which would reduce the overall range for the system by 1.5 dB.

When a ZPS is used instead of the CP, the ripples in the PSD can be reduced to zero with enough averaging. This is because the transmitted signal no longer has any structure; it is completely random. Figure 8.11 illustrates the ripples in the PSD for an MB-OFDM system that uses a CP and a ZPS. This figure shows that the ZPS will result in a PSD that is nearly flat, which corresponds to no reduction in transmit power (i.e., the system can achieve the maximum range possible).

8.4.6 System Parameters

The MB-OFDM system is capable of transmitting information at data rates of 53.3, 80, 106.7, 160, 200, 320, 400, and 480 Mbps. The system parameters for all the data rates are enumerated in Table 8.1. This system uses an OFDM modulation scheme with a total of 128 subcarriers, of which only 122 tones carry any energy. The energy-carrying tones are further divided into 100 data tones, 12 pilot tones, and 10 guard tones (5 on each side). These energy-carrying tones are all modulated using QPSK or dual-carrier modulation (DCM) (see Section 8.4.7). On each side of the OFDM symbol, the information placed on the guard tone data is generated by

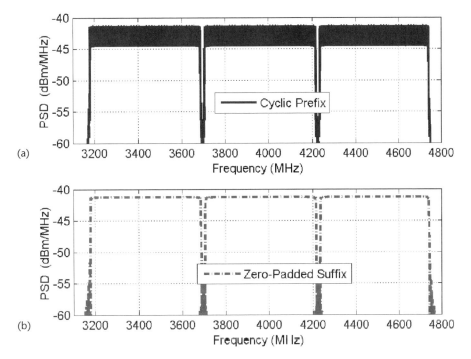

Figure 8.11 Power spectral density plots for an MB-OFDM system using (a) a CP, and (b) a ZPS.

replicating the information from the five outermost data-bearing tones [22]. The relationship between the power levels for the guard tones and data tones is assumed to be equal, even though in [13] it is left to the implementer to decide the exact ratio between the power levels.

The data rate of the system is varied through the use of FEC codes, frequency-domain spreading (FDS), and time-domain spreading (TDS). The various coding rates necessary to create the different data rates are generated by puncturing an industry standard $R = 1/3$, $K = 7$ convolutional code with generator polynomial $[133, 145, 175]_8$, where $[\cdot]_8$ refers to the octal representation of the polynomial. The exact puncturing patterns to generate the various coding rates are specified in [13]. FDS is obtained by forcing the input data into the IFFT to be conjugate symmetrical. The advantage of this type of spreading is that the output of the IFFT is always real, implying that only the real portion of the transmitter needs to be implemented or turned on. TDS is obtained by transmitting the OFDM symbol followed by a permutation of that same OFDM symbol.

8.4.7 Dual-Carrier Modulation

For the lower-data-rate modes, those of 200 Mbps and less, the MB-OFDM system exploits diversity in the frequency-selective channel through the use of FEC codes, FDS, and TDS. Unfortunately, for the higher-data-rate modes, those of 320 Mbps

228   Chapter 8

Parameter	Data Rate (Mbps)							
	53.3	80	106.7	160	200	320	400	480
Constellation	QPSK	QPSK	QPSK	QPSK	QPSK	DCM	DCM	DCM
FFT size	128	128	128	128	128	128	128	128
Code rate (R)	1/3	1/2	1/3	1/2	5/8	1/2	5/8	3/4
FDS	Yes	Yes	No	No	No	No	No	No
TDS	Yes	Yes	Yes	Yes	Yes	No	No	No
Number of data tones	100	100	100	100	100	100	100	100
Zero padded suffix (ns)	60.6	60.6	60.6	60.6	60.6	60.6	60.6	60.6
Symbol length (ns)	312.5	312.5	312.5	312.5	312.5	312.5	312.5	312.5
Multipath tolerance (ns)	70.1	70.1	70.1	70.1	70.1	70.1	70.1	70.1
Interleaver length (in symbols)	6	6	6	6	6	6	6	6

Table 8.1 *MB-OFDM System Parameters*

and greater, the only way to exploit channel diversity is through the use of FEC codes, especially when the data is modulated using QPSK. Any information that is unreliable due to deep fades in the channel can only be recovered using the FEC code.

In the MB-OFDM system, an additional form of channel diversity can be added by switching from a QPSK modulation to a DCM scheme. The additional diversity from the DCM scheme will result in an improvement in the overall range of the system, especially for the higher-data-rate modes. The DCM scheme introduces additional diversity by mapping four bits onto two 16-point constellations [23]. The symbols from the two 16-point constellations are then mapped onto tones that are separated by at least 200 MHz of bandwidth. A block diagram of the DCM technique is shown in Figure 8.12, and the mapping of bits onto the two 16-point constellations is shown in Figure 8.13.

It is easy to see how DCM exploits additional channel diversity. Suppose that the coded information on a single tone is unreliable due to the channel's experiencing a deep fade. In a system where QPSK is used instead of DCM, this information will

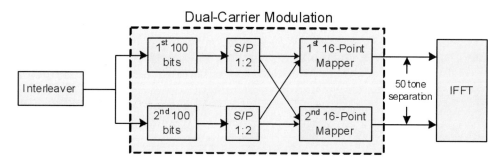

Figure 8.12 *Block diagram of the DCM.*

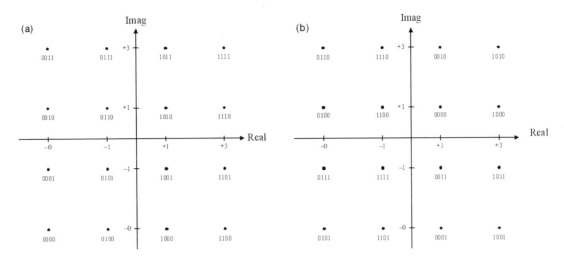

Figure 8.13 Two 16-point constellations used for the DCM technique.

have to be recovered using the strength of the FEC. However, in the DCM approach, the information can be recovered using a simple joint maximum a posteriori (MAP) decoder, which exploits the fact that the same information has been transmitted on two tones that are separated by at least 200 MHz. The probability that both channels carrying these tones will experience a deep fade is extremely small.

Since TDS and FDS are used for all rates below 320 Mbps, the diversity gains achieved by using DCM for 200 Mbps and less are minimal; therefore, the added complexity for DCM and the associated joint MAP decoder is not justified. Thus, DCM is only used for 320, 400, and 480 Mbps.

8.4.8 Link Budget Analysis for an AWGN Channel

The link budget for an MB-OFDM system with data rates of 106.7, 200, and 480 Mbps in an AWGN channel environment is shown in Table 8.2. In this analysis, an isotropic antenna (0 dBi antenna gain) and path-loss exponent of two have been assumed. In addition, the path-loss values are based on the geometric mean of the lower and upper frequency values; the geometric mean provides a more reasonable value for the expected path loss in the system.

The receiver noise figure value referenced in the link budget table is actually the noise figure referenced at the antenna, where the primary sources for noise are the LNA, mixer, transmit/receive switch, and preselect filter. In the analysis for the noise figure, it is assumed that the circuit impedance is 50 Ω, the voltage gain of the LNA is approximately 15 dB, and the voltage conversion gain of the mixer is approximately 10 dB. The total noise at the output of the LNA is 0.722×10^{-16} V^2/Hz, which includes the noise of the LNA and the input resistor. The total noise referred to the output of the LNA, including the referred mixer noise, is 0.722×10^{-16} V^2/Hz

230   Chapter 8

+ 0.1 × (8 × 10^{-9})^2 V^2/Hz = 0.786 × 10^{-16} V^2/Hz, where the second term in the addition is generated by the noise sources within the mixer. Thus, the overall noise figure for the analog front end is 10log$_{10}$ (7.86/2.56) = 4.9 dB. Including the losses associated with the preselect filter (1.1 dB) and the transmit/receive switch (0.6 dB), the overall noise figure referenced at the antenna is 6.6 dB. Losses due to ZPS overhead, front-end filtering, clipping at the DAC, ADC degradation, channel estimation, clock frequency mismatch, carrier-offset recovery, carrier-tracking, and so forth, have also been included in the implementation loss component of the link budget. The implementation loss value used in the link budget was derived from simulation results.

Table 8.2 shows that there is an excess link margin of 6 dB for an MB-OFDM system transmitting information at 106.7 Mbps at a distance of 10m, an excess link margin of 10.7 dB for a device transmitting information at 200 Mbps at a distance of 4m, and an excess link margin of 12.2 dB for a device transmitting information at 480 Mbps at a distance of 2m. The receiver sensitivity values for an MB-OFDM system operating in an AWGN channel environment and transmitting information at a rate of 106.7 Mbps, 200 Mbps, and 480 Mbps are –80.5 dBm, –77.2 dBm, and –73.2 dBm, respectively.

Parameter	Value	Value	Value
Information data rate (R_b)	106.7 Mbps	200 Mbps	480 Mbps
Average TX power (P_T)	~–10.3 dBm	~–10.3 dBm	~–10.3 dBm
TX antenna gain (G_T)	0 dBi	0 dBi	0 dBi
$f'_c = \sqrt{f_{min} f_{max}}$: geometric center frequency of waveform (f_{min} and f_{max} are the –10 dB edges of the waveform spectrum)	3,882 MHz	3,882 MHz	3,882 MHz
Path loss at 1m ($L_1 = 20\log_{10}([4\pi f'])/c$) $c = 3 \times 10^8$ m/s	44.2 dB	44.2 dB	44.2 dB
Path loss at dm ($L_2 = 20\log_{10}(d)$)	20 dB (d = 10m)	12 dB (d = 4m)	6 dB (d = 2m)
RX antenna gain (G_R)	0 dBi	0 dBi	0 dBi
RX power ($P_R = P_T + G_T + G_R + L_1 + L_2$ (dB))	~–74.5 dBm	~–66.5 dBm	~–60.5 dBm
Average noise power per bit ($N = -174 + 10 * \log_{10}(R_b)$)	~–93.6 dBm	~–91.0 dBm	~–87.2 dBm
RX noise figure referred to the antenna terminal (N_F)	6.6 dB	6.6 dB	6.6 dB
Average noise power per bit ($P_N = N + N_F$)	–87.0 dBm	~–84.4 dBm	–80.6 dBm
Required E_b/N_0 (S)	4.0 dB	4.7 dB	4.9 dB
Implementation loss (I)	2.5 dB	2.5 dB	2.5 dB
Link margin ($M = P_R - P_N - G_R - S - I$)	6.0 dB	10.7 dB	12.7 dB
Proposed minimum RX sensitivity level	~–80.5 dBm	~–77.2 dBm	~–73.2 dBm

Table 8.2 Link Budget Analysis for an AWGN Channel

8.4.9 System Performance in Multipath Channel Environments

The 90th percentile link success probability distance for an MB-OFDM system in various channel environments is listed in Table 8.3. The link success probability is defined as the percentage of channel realizations for which the system can successfully detect and demodulate a frame with a payload length of 1,024 octets at a PER of less than 8 percent. The channel models and the corresponding 100 channel realizations used for the simulations include four specific multipath environments: line-of-sight (LOS) with a range from 0m to 4m, NLOS with a range from 0m to 4m, NLOS with a range from 4m to 10m, and a worst case with an RMS delay spread of 25 ns as provided in [11]. Note that these channel realizations include a log-normal shadowing component with a standard deviation of 3 dB. Finally, the simulation results incorporate losses due to front-end filtering, clipping at the DAC, ADC degradation (four bits for 106.7/200 Mbps and five bits for 480 Mbps), packet acquisition, channel estimation, clock frequency mismatch (± 20 ppm at the TX and RX), carrier offset recovery, carrier tracking, and so forth.

In summary, the MB-OFDM system can support data rates of 106.7, 200, and 480 Mbps with a 90 percent link success probability at distances of approximately 11.5m, 7m, and 3.6m in various realistic multipath channel environments. The small variations in performance are primarily due to the effect of shadowing that has been incorporated in the 100 channel realizations corresponding to each of the four channel environments.

8.4.10 Packet Structure

Figure 8.14 shows the format of the MB-OFDM packet, which comprises three major components: the Physical Layer Convergence Protocol (PLCP) preamble, the PLCP header, and the PHY Service Data Unit (PSDU), which are listed in the order of transmission. The PLCP preamble is designed to aid the receiver in acquiring the packet, synchronizing to the packet, estimating the carrier-frequency offset, and estimating the effective channel. The second component of the packet is the PLCP header, which conveys the necessary information about the PHY and the MAC to aid in decoding the PSDU at the receiver. The PLCP header is composed of a PHY header, MAC header, header check sequence (HCS), tail bits, and Reed-Solomon

Channel Environment	Data Rate (Mbps)		
	106.7 (m)	200 (m)	480 (m)
AWGN	21.5	14.8	9.1
LOS: 0m to 4m	12.0	7.4	3.8
NLOS: 0m to 4m	11.4	7.1	3.5
NLOS: 4m to 10m	12.3	7.5	—
25 ns RMS delay spread	11.3	6.6	—

Table 8.3 Percent Link Success Probability Distance for an MB-OFDM System

232 Chapter 8

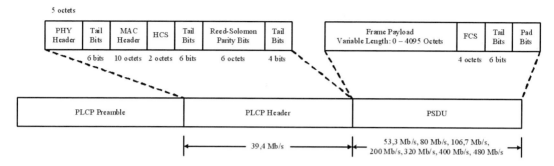

Figure 8.14 PLCP frame format.

(RS) parity bits. The tail bits are introduced to return the convolutional encoder to the "zero state," while the RS parity bits have been added to improve the robustness of the PLCP header (see Section 8.4.11).

The last component of the packet is the PSDU, which transfers the information via the frame payload. In addition to the frame payload, a frame check sequence (FCS), tail bits, and pad bits are added to the PSDU. The tail bits are again introduced to return the convolutional encoder to the "zero state," while the pad bits are introduced to align the packet boundaries with the boundary of the symbol interleaver. When the packet is transmitted, the PLCP preamble is sent first, followed by the PLCP header, at a data rate of 39.4 Mbps, and finally by the PSDU, at the desired data rate of 53.3, 80, 106.7, 160, 200, 320, 400, or 480 Mbps.

8.4.11 Robust PLCP Header

It is very likely that future versions of the MB-OFDM system will include advanced coding techniques, such as concatenated codes (convolutional inner code and a RS outer code) or low-density parity check (LDPC) codes. In fact, both academia and industry are already busy designing potential codes for the MB-OFDM system [24–26].

From a systems point of view, it is important that the PLCP header not be the weakest point of the packet (i.e., the PER should not be determined by the error rate of the header). The current format for the PLCP header is extremely robust because the PHY header (5 octets), MAC header (10 octets), and HCS (2 octets) are protected by an outer (23, 17) RS code and an inner R = 1/3, K = 7 convolutional code.

The (23, 17) RS code is generated from a systematic shortened (255, 249) RS code that has the following generator polynomial:

$$g(x) = \sum_{i=1}^{6}(x - \alpha^i)$$

where the RS code is defined over GF (256) with primitive polynomial $p(z) = z^8 + z^4 + z^3 + z^2 + 1$, and a is the root of $p(z)$. The outer code was designed to be a systematic

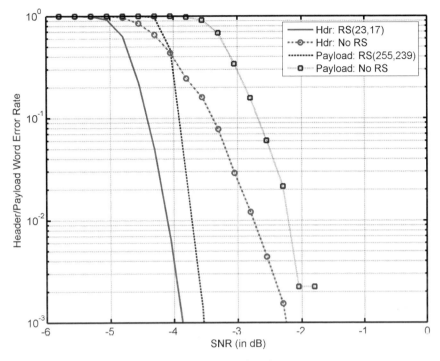

Figure 8.15 PLCP header performance gains with and without RS parity octets.

code to allow for receiver implementations that ignore the RS parity octets. Finally, the roots of the RS (23, 17) code are a subset of the roots of the RS (255, 239) code. Therefore, if an outer RS (255, 239) code is ever added to the payload section of the packet for an MB-OFDM system, the same encoder and decoder architecture can be used for the PLCP header and the frame payload.

Figure 8.15 illustrates the performance gains that can be achieved by adding RS parity octets to the PLCP header. At a PLCP header error rate of 1 percent, the performance gains are about 1.2 dB.

8.4.12 Multiple Access and Performance in the Presence of Other Piconets

Another important consideration in system design is the performance of the MB-OFDM device in the presence of other MB-OFDM interferers. The performance is primarily determined by the achievable signal-to-interference ratio (SIR), which is defined as

$$SIR = \left(\frac{P_{sig}}{P_{int}}\right)\left(\frac{W}{R}\right)$$

where P_{sig} is the power of the desired signal, P_{int} is the power of the interference, R is the information data rate, and W is the effective bandwidth of the transmitted signal. The first term in this equation, P_{sig}/P_{int}, is directly related to the distance between multiband devices, while the second term, W/R, is related to the bandwidth expansion factor, which indicates the processing gain available to suppress the interference. In an MB-OFDM system, the effective bandwidth is defined as

$$W = \frac{N_B \times N_{DT}}{T_S},$$

where N_B is the number of bands, N_{DT} is the number of data tones, and T_S is the symbol duration.

This equation shows that performance in the presence of MB-OFDM interference can be improved by ensuring there is a sufficient separation between the reference devices and the interfering devices. Unfortunately, in practice, this constraint is impossible to enforce. The other technique for improving the performance is to increase the bandwidth expansion factor or, equivalently, to make the effective bandwidth as large as possible.

There are several ways to achieve bandwidth expansion. The two most common and best-understood techniques are spreading (time domain and frequency domain) and FEC coding. In addition to these techniques, the MB-OFDM system uses a third, unique technique to obtain bandwidth expansion: time-frequency coding. A pictorial representation of how the information data is expanded in bandwidth is shown in Figure 8.16.

Essentially, the time-frequency codes specify the center frequency for the transmission of each OFDM symbol. These codes can be used to provide separation between piconets by assigning a unique time-frequency code to each piconet. Four of the time-frequency codes used in the MB-OFDM are listed in Table 8.4. These codes were designed to ensure that the average number of collisions between any two time-frequency codes is 1/3. In addition, the codes were also designed to ensure that the distribution of collisions should be as uniform as possible for all asynchronous shifts of the codes.

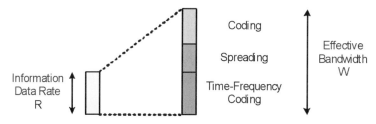

Figure 8.16 Pictorial representation of bandwidth expansion.

Channel (Piconet) Number	Time Frequency Codes					
1	1	2	3	1	2	3
2	1	3	2	1	3	2
3	1	1	2	2	3	3
4	1	1	3	3	2	2

Table 8.4 Time-Frequency Codes for Different Piconets

The performance of the MB-OFDM system using these four time-frequency codes in the presence of other MB-OFDM interferers is evaluated as follows. The distance between the transmitter and reference receiver for the reference link is set to a distance of d_{ref} = 0.707 of the 90 percent link success probability distance, where d_{ref} is the distance over which a system can operate without interferers. The separation distance at which an interfering device can be tolerated (i.e., the PER for the reference link is less than 8 percent) is obtained by averaging the performance over all combinations of the reference link and interferer link channel realizations for each channel environment, where the reference and interferer link channel realizations are specified in [27]. Finally, the d_{int}/d_{ref} values, where d_{int} is the distance between the reference receiver device and the interfering device and d_{ref} is the reference distance between the reference transmitter and the reference receiver, are shown in Table 8.5 as a function of the multipath channel environments and the number of interfering devices for an information data rate of 106.7 Mbps.

These simulations incorporated losses due to front-end filtering, clipping at the DAC, ADC degradation (four bits for 110 Mbps), multipath, packet acquisition, channel estimation, clock frequency mismatch (±20 ppm at the TX and RX), carrier offset recovery, carrier tracking, and so forth. In addition, the shadowing component was removed from both the reference and interfering links by normalizing each channel realization to have unit multipath energy.

These results show that a single interfering device can be brought within 3m of the reference without causing any disruptions in the reference link. Naturally, as more interfering devices are added to the scenario, the separation distance between the reference receiver device and the interfering device must also be increased. These results were obtained by exploiting the time-frequency codes, as well as the TDS techniques, and using symbol erasures when collisions were detected. More details about the receiver technique for erasing symbols can be found in [28].

Channel Environment	1 Interfering Device	2 Interfering Devices	3 Interfering Devices
CM1 (d_{int}/d_{ref})	0.40	1.18	1.45
CM2 (d_{int}/d_{ref})	0.40	1.24	1.47
CM3 (d_{int}/d_{ref})	0.40	1.21	1.46
CM4 (d_{int}/d_{ref})	0.40	1.53	1.85

Table 8.5 Performance in the Presence of Other MB-OFDM Interferers at a Data Rate of 106.7 Mbps

TFC	Band ID for TFC					
5	1	1	1	1	1	1
6	2	2	2	2	2	2
7	3	3	3	3	3	3

Table 8.6 Time-Frequency Codes for Improved Piconet Isolation

The isolation between piconets can be further enhanced by defining three additional time-frequency codes. These new time-frequency codes, which are listed in Table 8.6, are equivalent to frequency-division multiple access (FDMA). The isolation between piconets in an FDMA mode will depend on the roll-off of the receiver baseband filters, which in practice is a design choice for implementers.

This improved isolation does come at the expense of reduced range. Since only a single band is used, the effective average transmit power will be reduced by 4.8 dB as compared to a system using the time-frequency codes specified in Table 8.5. At most three overlapping piconets can operate simultaneously in this enhanced mode, primarily because only three FDMA channels can be defined.

Table 8.7 illustrates the distance d_{int} at which two interferers can be tolerated at a data rate of 106.7 Mbps, when the reference link is set to a distance of d_{ref} = 6m. In this test, the reference piconet is assigned band #2, while the interfering piconets are assigned bands #1 and #3. The baseband channel select filter is designed to provide an average adjacent channel rejection of approximately 15 dB. It is assumed that an interferer's signal always propagates through an AWGN channel.

Comparing the d_{int}/d_{ref} values in Table 8.7 with the results shown in Table 8.5 for the case when piconets use the full time-interleaving codes, it is clear that FDMA can provide a dramatic improvement in piconet isolation.

8.4.13 Complexity and Power Consumption

The total die size for the PHY solution is expected to be around 4.9 mm^2, where 3.0 mm^2 represents the component area of the analog/RF, and 1.9 mm^2 represents the digital baseband. These estimates assume a 90 nm CMOS technology node. If a 130 nm CMOS technology node is assumed, the total die size of the PHY solution is expected to be around 7.1 mm^2, with 3.3 mm^2 for the analog/RF and 3.8 mm^2 for the digital baseband. The digital portion of the PHY is expected to consume 295,000 gates. The major external components required are a preselect filter, balun, crystal oscillator, and voltage regulator.

Test Link Channel	d_{int}/d_{ref}
LOS: 0m to 4m	0.35
NLOS: 0m to 4m	0.33
NLOS: 4m to 10m	0.33

Table 8.7 Piconet Distance Separation When FDMA Is Used

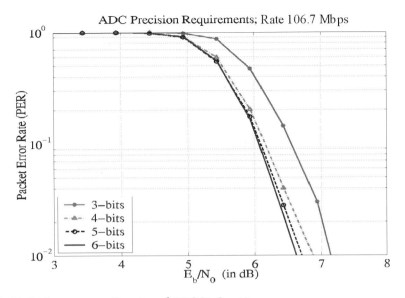

Figure 8.17 Performance as a Function of ADC Bit Precision.

The MB-OFDM system is specifically designed to be a low-complexity, CMOS-friendly solution. By limiting the constellation to QPSK and DCM, the resolution of the DACs and ADCs, as well as the internal precision in the digital baseband, is reduced. The bit precision of the ADC largely determines the power consumption of a UWB receiver. The performance degradation due to ADC bit precision was investigated in an AWGN environment for an information data rate of 106.7 Mbps. The PER performance of the MB-OFDM system is illustrated in Figure 8.17 as a function of the number of ADC bits. Based on Figure 8.17, we can conclude that the use of four bits at the front-end ADC is sufficient and results in less than 0.1 dB of degradation at a PER of 8 percent.

The estimated power consumption of an MB-OFDM implementation as a function of data rate is enumerated in Table 8.8. The power-consumption calculations are provided for both a 90 nm CMOS technology node and a 130 nm CMOS technology node. In addition, for the 90/130 nm process node, a supply voltage of 1.5V/1.8V is assumed for the RF/analog, except for the LNA, where a 2V supply is assumed. The digital baseband is assumed to have a supply voltage of 1.2V/1.3V for the 90-/130 nm process node, respectively, and a clock frequency of 132 MHz. Under these assumptions, the active power consumption for transmit, receive, clear channel assessment (CCA), and power-save modes are shown in Table 8.8.

8.4.14 Spectral Flexibility

A key challenge in UWB system design is that the regulations and spectrum allocation are currently only available in the United States. Other areas of the world (for example, Europe, Japan and Korea) are currently exploring the possibility of allocat-

Process Node	Rate (Mbps)	Active Transmit Power (mW)	Active Receive Power (mW)	Clear Channel Assessment (mW)	Power Save (Deep Sleep Mode) (μW)
90 nm	110	93	155	94	15
	200	93	169	94	15
	480	145	236	94	15
130 nm	110	117	205	117	18
	200	117	227	117	18
	480	180	323	117	18

Table 8.8 Power Consumption Numbers for an MB-OFDM System

ing spectrum for UWB devices. Regulations in other parts of the world have not been finalized, and the exact frequency allocation and emission limits may differ from those specified in the United States. Therefore, it is important that the system have sufficient spectral flexibility so that a single solution (chip and reference design) can be shipped throughout the world with only software changes required.

To illustrate the need for spectral flexibility, we consider the example of the radio astronomy bands in Japan. The Japanese regulatory authority appears to be strongly committed to protecting the radio astronomy bands. They propose that the emissions from UWB devices within these frequencies be limited to −64.3 dBm/MHz, which corresponds to a reduction in transmit energy of 23 dB over the emission level of −41.3 dBm/MHz specified by the FCC. Table 8.9 lists the radio astronomy bands that overlap with the UWB spectrum in the frequency range of 3.1 to 10.6 GHz [29].

Naturally, UWB devices that have the ability to sculpt their spectrum dynamically can introduce notches into the transmitted spectrum, thereby protecting the radio astronomy bands. By introducing notches, UWB proponents can remove any fears and doubts that radio astronomers may have about interference from UWB devices and thereby speed adoption and allocation of UWB spectrum in Japan.

Note that UWB devices effectively sacrifice bandwidth in order to protect the radio astronomy bands. Therefore, it is imperative to minimize the loss in bandwidth as this will affect performance. Ideally, the notches ought to have a sufficiently

Designation	Frequency Range (MHz)
S-band	3,260–3,267
S-band	3,332–3,339
S-band	3,345.8–3,352.5
C-band	4,800–4,990
C-band	4,990–5,000
C-band	6,650–6,675.2

Table 8.9 Japanese Radio Astronomy Bands

narrow bandwidth so as to null out only a particular radio astronomy band, and they should have a depth of at least 23 dB.

In summary, to ensure coexistence with other services, the ideal UWB device should have the ability to sculpt the spectrum while satisfying the following key criteria:

1. Reducing PSD in the band of interest. For example, in the radio astronomy band, this reduction should be 23 dB, corresponding to an emission level of –64.3 dBm/MHz.
2. Minimizing the loss of usable bandwidth, hence performance, needed to protect the radio astronomy bands.
3. Requiring minimal increase in transmitter complexity/hardware and no increase in receiver hardware.

As the MB-OFDM signal is constructed in the frequency domain, it is easier, both mathematically and computationally, to shape the spectrum of the transmitted signal. In addition to having the ability to turn on and off sub-bands, the MB-OFDM system has an intercarrier spacing of 4.125 MHz. Therefore, it is possible to shape the spectrum at a very coarse level, as well as at a very fine level. In an MB-OFDM system, a single 7 MHz-wide radio astronomy band would overlap with at most three tones of the MB-OFDM system. A number of techniques can be used to shape the spectrum of the MB-OFDM signal [30].

8.4.14.1 Tone Nulling

The most common technique for creating a notch in the frequency domain is to zero out tones that overlap with the radio astronomy band. The advantage of this technique is that there is no increase in complexity at the transmitter. Additionally, the receiver does not require any prior knowledge of the notch. At the receiver, tones carrying no information will look like a deep fade in the channel. Since the receiver is not able to differentiate between these two phenomena, no prior information needs to be communicated to the receiver in order to compensate for the tones carrying no information.

The transmitted OFDM signal is constructed using an inverse discrete Fourier transform (IDFT). As a rectangular window is applied to the data, each tone has a wider-than-expected spectrum, where the spectrum has the shape of a sinc function. Although the sinc function has zero crossings at each of the tone locations, zeroing out only a few tones results in a shallow notch. For example, to obtain a notch with a depth of 23 dB for the radio astronomy band, a total of 29 tones need to be zeroed out. This corresponds to a total loss of 120 MHz of bandwidth (i.e., nearly 8 percent of the spectrum for a three-band system). Using any fewer tones does not produce the desired notch depth. This is illustrated in Figure 8.18, where a total of 11 tones (~45 MHz of spectrum) are zeroed out to obtain just 15 dB of suppression.

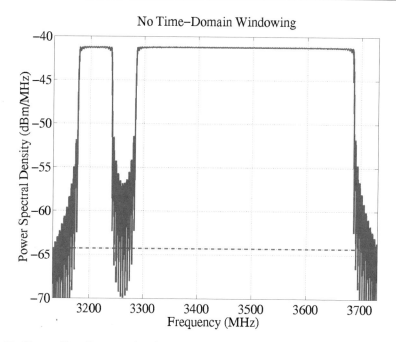

Figure 8.18 Tone nulling for spectral sculpting of the MB-OFDM system.

8.4.14.2 Dummy Tone Insertion

A deeper notch can be obtained by inserting data-specific dummy tones on either side of the victim band [31]. The spectral leakage, which is the root cause of the wider notches, is dependent on the data that is transmitted on the nonzero tones. By accounting for the contribution due to the nonzero tones, dummy tones can be introduced to minimize the spectral leakage and thereby increase the depth of the notch.

We now describe the method to determine the value of the dummy tones as a function of the information transmitted in the data tones for each OFDM symbol. The spectral leakage on each of the subcarriers due to the application of the N-point IDFT operation is given as

$$W(f) = \frac{1}{\sqrt{N}} \sum_{n=0}^{N-1} \exp\left(\frac{-j2\pi nf}{N}\right)$$

The frequency-domain response of each transmitted OFDM symbol is

$$S(f) = \sum_{k=0}^{N-1} S(k) W(f-k)$$

in which $S(k)$ refers to the data/pilot information transmitted on the kth subcarrier. Let the frequency-domain response be sampled at frequencies that are 1/4 of the intercarrier spacing apart. Then

$$\underline{b} = P\underline{s}$$

where $\underline{b}$ is an $L \times 1$ vector that represents the frequency-domain spectrum at frequencies within the radio astronomy band, $\underline{s}$ is an $N \times 1$ vector with the data/pilot/null tones, and P is an $L \times N$ matrix derived by picking L rows from a circulant matrix computed by sampling $W(f)$. The $(l,k)^{\text{th}}$ entry of the matrix P is given below:

$$P_{l,k} = \sum_{n=0}^{N-1} \exp\left(\frac{-j2\pi n\{f(l)-k\}}{N}\right)$$

where $f(l)$ corresponds to the frequency of the radio astronomy band at the lth sampling point. The data-specific dummy tones are then determined by minimizing the cost function:

$$\underline{\hat{x}} = \arg\min_{x} \|P_1 \underline{x} + \underline{b}\|_2^2$$

where $\underline{x}$ is an $M \times 1$ vector representing the data to be inserted in the M dummy tones, and P_1 is an $L \times M$ submatrix derived by picking M columns from the matrix P corresponding to the dummy tone locations. The least-squares solution for the dummy tones is

$$\underline{\hat{x}} = \left(P_1^H P_1\right)^{-1} P_1^H \underline{b}$$
$$= \left(P_1^H P_1\right)^{-1} P_1^H P\underline{s}$$

Using this method, a notch with a depth of 23 dB in the frequency band of 3,260–3,267 MHz can be generated by using a total of just five tones: three zero tones, which overlap with the band of interest, and a data-specific dummy tone on each edge of the victim band. The dummy tone is loaded such that it cancels out the spectral leakage into the radio astronomy band. The resulting PSD of the transmitted signal is illustrated in Figure 8.19.

The advantage of this technique is that only a total of five tones need to be sacrificed to insert a 23 dB notch in the 3,260–3,267 band. This corresponds to a loss of about 20 MHz of spectrum (i.e., less than 1.3 percent of the spectrum for a three-

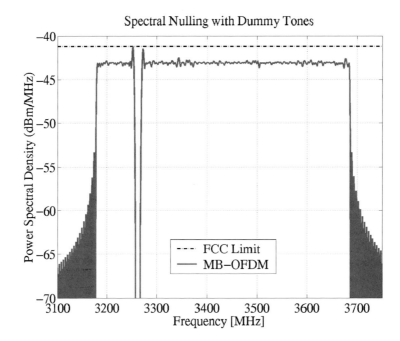

Figure 8.19 Spectral nulling of a radio astronomy band for an MB-OFDM system.

band system). Note that these five tones are sufficient to introduce even deeper notches.

However, a severe limitation of this technique is that it increases the ripple in the PSD and specifically causes peaking at the band edges. To meet the FCC specifications, the PSD of the UWB system should be below the −41.25 dBm/MHz level. The resolution bandwidth for the FCC compliance measurement for average PSD is 1 MHz, and the averaging time is about 100 ms. This necessitates a transmit power back-off of nearly 1.8 dB and, hence, a loss in performance or, equivalently, range. Note that a 1.8 dB back-off in transmit power is equivalent to losing nearly one-third of the usable spectrum. This is a very serious drawback of this technique.

8.4.14.3 Constrained Dummy Tone Insertion

To avoid peaking in the PSD of the transmitted MB-OFDM signal, we propose a modification to the way in which the dummy tones are derived. Although insertion of the dummy tone reduces the spectral emissions in the band of interest (e.g., the radio astronomy band), there is no guarantee that it will not increase the emissions outside this band. Specifically, as the magnitude of the dummy tones is not constrained at frequencies outside of the radio astronomy band, it introduces peaking at the edges of the notch.

Ideally, we need to constrain the peak spectral content in the UWB frequencies in addition to minimizing the emissions in the radio astronomy band. However, introducing such a constraint makes the optimization intractable and does not result in a closed form solution. Hence, to mitigate the ripples in the PSD, we propose the following modification to the cost function. We add a regularization term that constrains the amount of energy placed on the dummy tones and indirectly reduces the emissions in the UWB frequencies. This is mathematically represented as

$$\underline{\hat{x}}^C = \arg\min_x \left\{ \left\| P_1 \underline{x} + \underline{b} \right\|_2^2 + \lambda \left\| \underline{x} \right\|_2^2 \right\}$$

where λ is a weighting coefficient for the regularization term. The least-squares solution then turns out to be

$$\underline{\hat{x}}^C = \left(P_1^H P_1 + \lambda \right)^{-1} P_1^H \underline{b}$$
$$= \left(P_1^H P_1 + \lambda \right)^{-1} P_1^H P \underline{s}$$

The efficacy of this technique is illustrated in the PSD plot of Figure 8.20, in which a λ value of 16 has been used. We are able to introduce a 23 dB notch with minimal ripple in the passband. Additionally, the value of λ can be predetermined, and the matrix inverse and the matrix product terms can be precomputed.

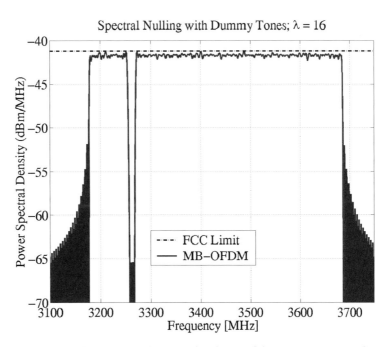

Figure 8.20 *Constrained dummy tones for spectral sculpting of the MB-OFDM signal.*

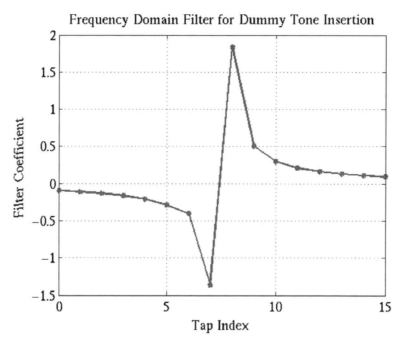

Figure 8.21 Impulse response of a dummy tone computation filter.

8.4.14.4 Complexity

This technique requires the computation of only two values, which are to be loaded onto the dummy tones, for each OFDM symbol transmitted with a notch. Since the notch only exists for OFDM symbols transmitted on band #1, this computation needs to be performed, on average, for only one-third of the transmitted OFDM symbols (i.e., once every 0.9375 s). The two dummy tones can be computed using a finite-tap frequency-domain filter on the data and pilot tones. Since the data and pilot tones are from a QPSK constellation, the filtering requires only addition and subtraction operations. As an example, a 16-tap frequency-domain filter with real coefficients can be used to compute the data-specific information for the two dummy tones. In terms of computational complexity, this filtering would require a total of 68.5 mega addition operations per second. The response of the frequency-domain filter is illustrated in Figure 8.21.

8.5 Conclusion

The FCC created a great opportunity in February 2002 when 7,500 MHz of spectrum were allocated for unlicensed use with commercial UWB devices. The IEEE 802.15.3a task group has developed a channel model to estimate the performance of UWB systems in real-world environments. This has allowed designers to develop a specification, based on MB-OFDM, which meets the stringent market requirements: hundreds of megabits per second at low power and low cost. Several key organizations (MBOA, WiMedia, Wireless USB) have selected this design for their applications. OFDM already enjoys an outstanding record with other standards organizations, such as Asymmetric Digital Subscriber Line (ADSL), IEEE 802.11a, IEEE 802.11g, and IEEE 802.16a. In addition, OFDM was adopted for digital audio and terrestrial video broadcast in both Europe and Japan. The choice by the UWB industry is based on the facts detailed here that show how MB-OFDM presents a very good technical solution for the diverse set of high-performance, short-range applications eagerly anticipated by the consumer electronics, personal computing, and mobile applications industries.

8.6 Acknowledgments

The authors would like to thank Anand Dabak, Ranjit Gharpurey, and Jerry Lin for their efforts in developing the MB-OFDM proposal and Ranjit Gharpurey for helping to develop the architecture for the synthesizer described in Section 8.4.3 for providing the simulations results for the frequency synthesis circuit. Finally, the authors would like to acknowledge the contributions and support of all members in the MB-OFDM Alliance and WiMedia Alliance that have made it possible for this design to become a market reality.

References

1. Federal Communications Commission. "First Report & Order, Revision of Part 15 of the Commission's Rules Regarding Ultra-Wideband Transmission Systems." ET Docket 98-153, February 14, 2002.
2. FCC 47 C.F.R. 15.5(b).
3. FCC 47 C.F.R., 1.907, 2.1.
4. Win, M. Z., and R. A. Scholtz. "On the Robustness of Ultrawide Band Signals in Dense Multipath Environments." *IEEE Comm. Letters* 2, no. 2 (February 1998).
5. Somayazulu, V. S., J. R. Foerster, and S. Roy. "Design Challenges for Very High Data Rate UWB Systems." *Proc. Asilomar Conf. on Systems, Signals and Comp.* (November 2002): 717–721.
6. Welborn, Matt, et al. "XtremeSpectrum CFP Proposal." IEEE P802.15-03/334r3-TG3a (September 2003).
7. Saleh, A., and R. Valenzuela. "A Statistical Model for Indoor Multipath Propagation." *IEEE Journal on Selected Areas in Communications* 5, no. 2 (February 1987): 128–137.
8. Hashemi, H. "Impulse Response Modeling of Indoor Radio Propagation Channels," *IEEE Journal on Selected Areas in Communications* 11, no. 7 (September 1993): 967–978.

9. Cassioli, D., M. Z. Win, and A. F. Molisch. "The Ultra-Wide Bandwidth Indoor Channel—From Statistical Model to Simulations." *IEEE Journal on Selected Areas in Communications* 20, no. 6 (August 2002): 1247–1257.

10. See http://ultra.usc.edu/New_Site/database.html.

11. Foerster, J., ed. "Channel Modeling Sub-committee Report Final." IEEE 802.15-02/490, November 2002, at http://ieee802.org/15.

12. Balakrishnan, J., A. Dabak, S. Lingam, and A. Batra. "Complexity and Performance Analysis of a DS-CDMA UWB System." IEEE P802.15-03/388r2 (September 2003).

13. *High Rate Ultra Wideband PHY and MAC Standard, 1st ed.* Geneva: ECMA, December 2005.

14. Batra, A., et al. "TI Physical Layer Proposal for IEEE 802.15 Task Group 3a." IEEE P802.15-03/142r2-TG3a (March 2003).

15. Batra, A., et al. "Multiband OFDM Physical Layer Proposal." IEEE P802.15-03/268r0-TG3a (July 2003).

16. Balakrishnan, J., A. Batra, and A. Dabak. "A Multiband OFDM System for UWB Communication." *Proc. IEEE Conf. on Ultra Wideband Systems and Technologies*, Reston, Virginia (November 2003): 354–358.

17. Batra, A., J. Balakrishnan, and A. Dabak. "Multiband OFDM: A New Approach for UWB." *Proceedings of 2004 IEEE International Symposium on Circuits and Systems* 5 (May 2004): 365–368.

18. Batra, A., J. Balakrishnan, et al. "Multiband OFDM Physical Layer Proposal for IEEE 802.15 Task Group 3a." IEEE P802.15-04/493r1-TG3a (September 2004).

19. Batra, A., J. Balakrishnan, G. R. Aiello, J. R. Foerster, and A. Dabak. "Design of a Multiband OFDM System for Realistic UWB Channel Environments." *IEEE Transactions on Microwave Theory and Techniques* 52, no. 9, part 1 (September 2004): 2123–2138.

20. Bingham, J. A. C. "Multicarrier Modulation for Data Transmission: An Idea Whose Time Has Come." *IEEE Communication Magazine* 28, no. 5 (May 1990): 5–14.

21. Muquet, B., Z. Wang, G. B. Giannakis, M. de Courville, and P. Duhamel. "Cyclic Prefix or Zero Padding for Wireless Multicarrier Transmission?" *IEEE Transactions on Communications* 50, no. 12 (December 2002): 2136–2148.

22. Batra, A., J. Balakrishnan, A. Dabak, and S. Lingam. "What Is Really Fundamental?" IEEE P802.15-04/533r0-TG3a (September 2004).

23. Private conversations with Srinath Hosur, Andy Molisch, Eric Ojard, and Yossi Erlich.

24. Rashi, Y., E. Sharon, and S. Litsvn. "LDPC for TFI-OFDM PHY." IEEE 802.15-03/353r0-TG3a (September 2003).

25. Png, K.-B., and X. Peng. "Performance Studies of a Multiband OFDM System Using a Simplified LDPC Code." *2004 Joint IEEE Conf. UWBST and IWUWBS*, Kyoto, Japan (May 2004): 376–380.

26. Kim, S.-M., J. Tang, and K. K. Parhi. "Quasi-Cyclic Low-Density Parity-Check Coded Multiband OFDM UWB Systems." *Proceedings IEEE International Circuits and Systems Symposium* 5 (May 2005): 65–68.

27. Roberts, R., K. Siwiak, and J. Ellis, eds. "P802.15.3a Alt PHY Selection Criteria." IEEE P802.15-03/31r6-TG3a (January 2003), at http://ieee802.org/15.

28. Park, S. Y., G. Shor, and Y. S. Kim. "Interference Resilient Transmission Scheme for Multiband OFDM System in UWB Channels." *Proceedings of 2004 IEEE International Symposium on Circuits and Systems* 5 (May 2004): 373–376.

29. Radio Astronomy Frequency Allocations, at www.ukaranet.org.uk/basics/frequency_allocation.htm.

30. Batra, A., S. Lingam, and J. Balakrishnan. "Multiband OFDM: A Cognitive Radio for UWB." *Proceedings of 2006 IEEE Symposium on Circuits and Systems*, Island of Kos, Greece, 2006.

31. Yamaguchi, H. "Active Interference Cancellation Technique for MB-OFDM Cognitive Radio." *European Microwave Week*, Amsterdam, Netherlands, 2004.

9

MAC Designs for UWB Systems
by Larry Taylor

9.1 Introduction

UWB is a relatively new technology for communications systems, and there are very few completed designs or standards. However, two important specifications—Certified Wireless USB (WUSB) and WiMedia—have been completed, and they are described in this chapter. Although these two specifications have been developed to share the same UWB Physical Layer, they address very different architectural goals.

9.1.1 Communications Model

Figure 9.1 depicts the lower layers of the standard OSI Reference Model, together with the partitioning of the Data Link Layer as used to describe the IEEE 802 family of standards. Most local area network (LAN) standards are described using this model, including 802.11, the most successful wireless LAN (WLAN) standard. An extension of the WLAN architecture for short-range wireless systems is the 802.15 wireless personal area network (WPAN) with a nominal extent of 10m centered on a person or system. The 802.15 architecture defines a centrally controlled piconet,

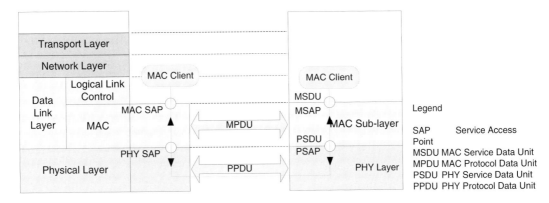

Figure 9.1 Communications model.

and some of its concepts are adopted in the WUSB architecture. Such architectures are less suited to dynamic communications environments, and the WiMedia MAC diverges from the piconet concept, as we shall see in Section 9.3.

The MAC architecture defines a communications channel over the Physical layer connecting two MAC entities. MAC clients, which may be upper-layer protocols such as Internet Protocol (IP) or applications directly accessing the MAC, communicate with the MAC through the MAC service access point (MAC SAP). Associated with the MAC SAP is an address, usually globally unique, such as an extended unique identifier (EUI-48 or EUI-64). MAC entities execute the MAC protocol by exchanging MAC protocol data units (MPDUs) using an addressing mechanism suited to the MAC architecture. The MPDUs are in turn encapsulated in PHY protocol data units (PPDUs).

The MAC sublayer interacts directly with the Physical layer to regulate access to the communications channel and provide a data-transfer service for the MAC client. The MAC architecture is constructed to exploit the characteristics of the PHY and, of course, is also constrained by inherent PHY-specific limitations.

9.1.2 UWB PHY Characteristics

The most important properties and characteristics of a UWB PHY from the MAC design perspective are

- Half-duplex communications
- Shared communications channel
- High sampling rate
- High capacity
- High instantaneous data rate
- Limited range
- Time-domain waveform properties

The shared, half-duplex communications channel results directly from the wide bandwidth of the radio subsystem. The implications for the MAC design include considerations of finite transmit-receive turnaround time for efficiency in protocol design and fairness in the medium access rules.

The high sampling rate means that the MAC should minimize time spent listening to the channel in order to conserve energy, especially for battery-operated systems.

High capacity and high data rate are related.

$$C = W \log_2(1 + SNR) \tag{9.1}$$

Shannon's Law states that the capacity, C, of a communications channel is proportional to its bandwidth, W, and the signal-to-noise ratio of the received signal, as shown in (9.1). For UWB signals, the capacity is large by reason of a large occupied bandwidth.

Capacity translates into high data rate as either a high aggregate data rate from a large population of simultaneously operating devices or a high instantaneous data rate from each of a small number of simultaneously operating devices. The higher the data rate, the greater the amount of signal energy that needs to be captured to demodulate the signal successfully; hence, the range will be less than it would be for a lower-data-rate signal transmitted at the same power. For high-data-rate systems, it is essential that the MAC provide high efficiency as even small amounts of unusable time equate to significant loss in potential data transfer.

Many of the attractive features of UWB signals are due to the time-domain nature of the signal transmission. Early UWB systems generated simple, short pulses with wide spectral occupancy and simple modulation. More recent waveforms use higher-order modulation schemes on constrained bandwidth to enable efficient channelization. In all cases, the ability to capture the leading wavefront of a UWB signal enables time of flight from the transmitter to the receiver to be determined with accuracy proportional to the occupied bandwidth, which in turn provides support for distance estimation or ranging.

Together, these characteristics of shared half-duplex channels, namely, high capacity, implying high data rate, and time-domain signals enabling ranging, have a strong influence on the architecture of MAC designs for UWB systems.

9.1.3 Function of the MAC Sublayer

The main function of the MAC sublayer for a shared communications channel is to mediate access to the channel resource. To understand the design issues the UWB MAC designer faces, it is first necessary to investigate possible deployment topologies.

Figures 9.2 and 9.3 represent a cluster of UWB devices A to D where cluster is used to mean "in close physical proximity." The circles represent the idealized radio range of each device. Devices are within mutual range if their range circles intersect. As shown in these diagrams, Device A is within radio range of each of the other devices. Devices B and C are within mutual range and in range of Device A. Device D is only within range of Device A.

This partially connected network is characteristic of the topology UWB MAC designs must deal with, primarily because of the elevated data rates being used and the consequent limited communications range.

UWB MAC systems can be grouped into two classes as shown in Figure 9.3. The class on the left is a centrally controlled architecture where each device communicates with a designated cluster controller (Device A). The fact that some of the devices are out of mutual communications range is unimportant as long as all of the devices in the cluster are within range of the cluster controller.

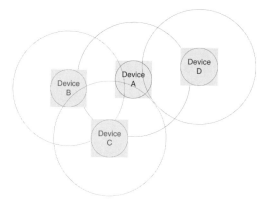

Figure 9.2 UWB device arrangement.

The class on the right is a distributed control cluster where each device may communicate directly with any neighboring device within range.

The design of a MAC, in addition to being constrained by the characteristics of the PHY, must address the goals of expected applications. High-performance communications systems are expected to offer services capable of dealing with very demanding applications usually described as streaming media applications. These applications may need sustained, elevated data rates for their operation. Other applications demand tight limits on transit delay and delay jitter. Yet other applications require high data-transfer rates with low loss.

Radio communications systems suffer from impairments in the channel, such as noise, multipath effects, and interference from both collocated systems using the same channel and energy from adjacent channel systems. These impairments combine to cause data loss, and the operation of the MAC protocol must both take into account the imperfect nature of the communications channel for its own communications and avoid generating interference to others.

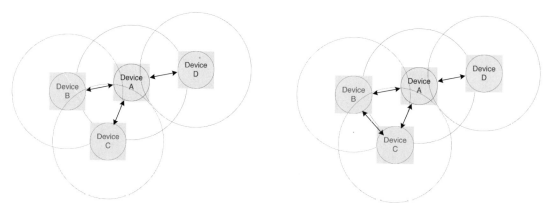

Figure 9.3 Central and distributed arrangements.

9.1.3.1 Address Space

As mentioned earlier, the MAC SAP is usually addressed via a universally unique identifier such as a EUI-48 or EUI-64 since the MAC client may be any of a very large number of systems and applications. However, as can be seen from the brief analysis of possible topologies, and taking into account the short range of the high-data-rate communications, the number of systems that may be addressed within the MAC is small compared to these universal address spaces.

Since the MPDUs carry two addresses (source and destination), the protocol overhead can be reduced by using smaller address fields. The effect on a specific MAC depends on its architecture:

- A centralized architecture, such as WUSB, has a cluster controller that can allocate device addresses (USB has a limit of 127 active devices).
- A distributed architecture, such as WiMedia, has no control over the neighborhood and, consequently, must provide means to ensure that dynamically assigned addresses are unambiguous in at least the two-hop neighborhood.

Other addressing issues will be discussed later in the chapter.

9.1.3.2 Physical Layer Assumptions

A common PHY is assumed in the description of each of the UWB MAC examples later in the chapter. This PHY provides the following:

- Estimation of medium activity
- Error detection for PHY and MAC header structures and frame payload
- Frame transmission and reception with precise timing of both start and end of frame
- Burst transmission with reduced protocol overhead
- Precise time references to support range estimation

Figure 9.4 shows the structure of a PHY PDU.

The preamble is a predefined symbol sequence used for frame detection and provides the precise start of frame timing information required for the MAC protocols and range estimation. The PHY Layer Convergence Protocol (PLCP) header,

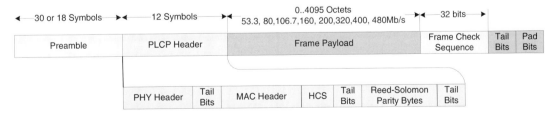

Figure 9.4 PHY PDU structure.

including both the MAC and PHY headers, is protected by a strong header check sequence (HCS). Note that since the MAC header is carried and protected in the PLCP header, all protocols using this PHY PDU structure share the same MAC header parameters.

PHY-specific parameters, such as frame payload data rate and length, are carried in the PHY header, which is transmitted at a special highly robust data rate and is captured and interpreted by all recipient devices. The MAC header is also common and carries information about frame addressing, as well as various frame control and sequence information. The frame payload is protected by a separate check sequence and may be transmitted at a higher data rate supported by both the source and destination devices. The PHY PDU may be padded to the number of octets needed to fill an integer number of PHY symbols.

The MAC controls the values of the PHY parameters, including the PHY channel, and the start of frame timing. Burst mode is a PHY facility, however, and its timing is controlled by the PHY.

With these assumptions defined, we can investigate the example UWB MACs for centralized and distributed control.

9.2 Centralized Control Architecture

WUSB is a specification published by the USB Implementer's Forum (www.usb.org/developers/wusb) that builds on the highly successful wired USB architecture (see USB 2.0 specification www.usb.org/developers/docs), which supports peripheral device interconnections to a USB host system at up to 480 Mbps. In this description of the WUSB MAC architecture, an understanding of wired USB is assumed, although not required, and below we provide a summary of the key points as needed to understand UWB MAC issues.

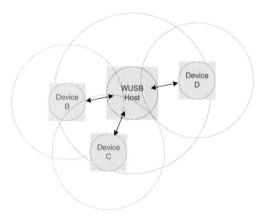

Figure 9.5 WUSB cluster.

WUSB uses a centralized control architecture with host and device roles (see Figure 9.5), where the WUSB host provides all system control, timing, and data-flow control. Before communication can take place, a WUSB device must associate with a WUSB host. The WUSB host authenticates the device before providing it with an address to use in subsequent communications. All data exchanges between the WUSB host and its associated devices are secure, being either encrypted or simply authenticated. Except where it is not possible owing to the wireless medium, the WUSB architecture follows the wired USB architectural model and allows the large existing base of software to operate in the new wireless environment with little or no modification.

9.2.1 Wired USB Model

The wired USB model is a centralized control architecture bus operating over a four-wire cable with a differential data pair, power, and ground. The USB host is the master and provides all system timing and control. It is not necessary to describe the USB protocol in detail here, but a basic understanding of the electrical operation, device connection, enumeration, data transfer, and disconnection is helpful in understanding issues for the WUSB architecture.

Communication over the wired USB bus is performed by driving the differential signal pair according to a strict timing protocol using non-return-to-zero inverted (NRZI) data encoding to provide sufficient transitions on the data bus to recover clock timing accurately. The same pair is also sensed by single-ended receivers to detect the connection and disconnection of devices to the bus. USB defines three device speeds, low speed, full speed, and high speed. The speed of a device is recognized when it connects to the bus by the effect it has on the single-ended voltages on the differential data lines. The USB host can provide power to the devices over the bus voltage lines. Communication is secure by reason of the cabled link between the host and the device.

Because wired USB is a bus topology, special functions are defined for expanding the bus via a local branching device called a hub. A hub supports one or more devices and passes signals from the host to each attached device. The hub generates bus timing for its attached devices and responds to polls from the host for information about new device attachments or other similar events.

When a device is connected to the bus, the host (or hub) detects the new device and resets the bus. The reset operation allows the host to enumerate the device by polling default addresses (device address and endpoint within the device) to obtain configuration information, establish a communications schedule to satisfy all of the connected device requirements, and compute a power budget. If the host cannot satisfy the device's data or power requirements, the device is not configured. The device has only two major states, unconfigured or configured.

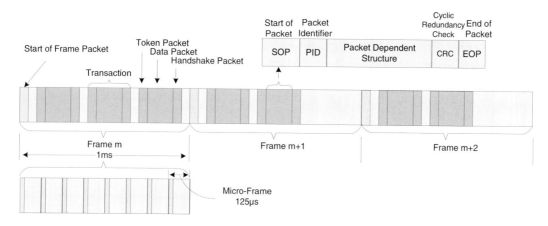

Figure 9.6 Wired USB communication structure.

Communications take place on the bus in transactions scheduled within 1 ms frames (or 125 μs microframes for high-speed transactions) (see Figure 9.6). Figure 9.7 shows a simplified view of the USB transaction. The host transmits a token packet on the bus indicating the device, the endpoint within the device, and the type of transfer. The data follows, either from the host to the addressed device or from the addressed device to the host. The transaction completes with a handshake packet (Hsk in Figure 9.6) indicating the status of the data transfer. Each packet within the transaction is preceded by a short synchronization preamble (8 or 32 bits) and followed by a short end-of-packet sequence. The data is protected by a cyclic redundancy check (CRC) field, and there is a short interpacket gap. Wired USB defines

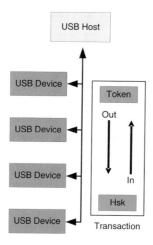

Figure 9.7 Wired USB architectural model.

four types of endpoint (control, interrupt, bulk, and isochronous) to support various data needs, including asynchronous and isochronous data flows.

Signal timing is very precise, and the bus is highly reliable with a bit error rate of the order 1×10^{-11}. The synchronization preamble, end-of-packet sequence, and minimum interpacket gap are all short, ensuring a low fixed overhead for the transactions. The schedule of transactions repeats each millisecond such that low-latency devices are serviced with good response time and high-capacity devices receive regular service to minimize buffering requirements.

9.2.2 WUSB Model

The WUSB model is quite similar to the wired USB model in many respects. As in wired USB, there is a single host and one or more devices in a WUSB cluster. The host retains full control of the operation of the bus and provides all bus timing and flow control.

However, the wireless medium introduces some significant differences that give rise to the following issues for the wireless MAC:

- A WUSB device may be within radio range of, and able to communicate with, multiple WUSB hosts. Hence, the device connection states must be expanded to include something equivalent to connecting a wired USB device to the wired bus.
- Since there is no electric connection between the members of the WUSB cluster, connection and disconnection must be handled differently.
- The wired USB cable provides strong security with little chance of outside eavesdropping and practically no chance of data modification. The radio medium, on the other hand, is a broadcast medium, and any device within range can capture the radio signal and attempt to decode it. Various security threats are exposed, including replaying previous data and data modification.
- Whereas the electrical signals on the wired USB cable can be easily detected, and changes between transmit and receive states have very small delays, the radio receiver requires significantly longer times for signal capture and synchronization and for changes in receive and transmit states, giving rise to longer interpacket gaps.
- The raw error rate of the wireless medium is very high compared to that of the wired USB cable such that the wireless MAC must include checks and retransmissions to achieve acceptable packet error rates. These additional measures add to the protocol overhead and increase latency and jitter.

WUSB device functions are almost the same as wired USB device functions, allowing existing USB software to run over WUSB in most cases, except for isochronous data flows. The system software model of endpoints and data flows is similarly close to the wired USB model except that a secure context must first be established before communications are allowed to take place. However, the interface with the wireless "bus" is significantly different from that of the wired USB bus.

9.2.2.1 WUSB State Model

A WUSB device transits the states shown in Figure 9.8 (the device and host must have previously been associated, exchanged a shared secret, and saved a "Connection ID" generated by the host). The device by default resides in the unconnected state. A device sends a "connect" request to a WUSB host using a reserved "unassigned" device address and monitors the WUSB host's responses for permission to begin an authentication procedure.

If granted, the device enters the connected, but unauthenticated, state and uses a device address assigned by the host.

A four-way authentication procedure is executed to authenticate the device and host pair and to exchange session keys to be used to secure subsequent data communications between them. If the authentication procedure succeeds, the device enters the connected and authenticated state and is fully operational. The device remembers the security context established and may reuse it when reassociating with the same host at a later time.

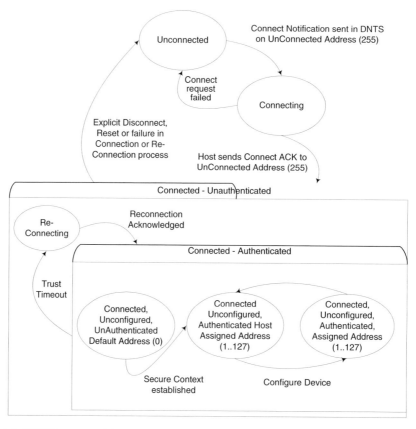

Figure 9.8 WUSB state model.

9.2.3 WUSB Device Addressing

Although the WUSB architecture only permits up to 127 WUSB devices to be connected to a WUSB host, WUSB frame addressing is based on locally assigned 16 bit device addresses. The WUSB architecture requires WUSB host control over all device address assignments.

To permit new devices to become members of a WUSB cluster, the WUSB architecture reserves certain values in the 16 bit address space for specific purposes. (See Table 9.1.)

The general rule for WUSB device addressing is that the WUSB host assigns a WUSB device an address except when the WUSB device is in the unconnected state, in which the WUSB device may only use the unconnected device address (255).

A WUSB host may reset a WUSB device into a default state—a condition in which the WUSB device remains a member of the WUSB cluster but loses any previous confirmation data such as endpoint states—by assigning address zero via a control operation.

The range 128 to 254 is used to select a cluster broadcast address, which is used for all clusterwide communications and also for temporary WUSB device communications before the WUSB device has been authenticated.

The WUSB host selects its device address to be compatible with other WUSB hosts or non-WUSB devices sharing the same channel. Coexistence of multiple systems and protocols is discussed in Section 9.2.7.

Normal cluster operation uses device addresses in the range from 1 to 127.

9.2.4 WUSB Frame Structure

Figure 9.9 expands the PHY PDU described earlier. The MAC header fields provide control, frame type, and addressing information.

9.2.4.1 Frame Control

Each MAC frame is transmitted with a parameter set to assist in the reception and processing of the frame. The frame control field holds the majority of frame-

Address Value	Description
0	Default Address–WUSB device is in the Default state
1..127	Host assigned WUSB device address
128..254	Cluster Broadcast Address range. Temporary WUSB device address during authentication procedure
255	Un-connected device address. Used by WUSB device in the connection procedure
256..65279	Host Device Address permitted range
65280-65535	Unused

Table 9.1 WUSB Address Ranges

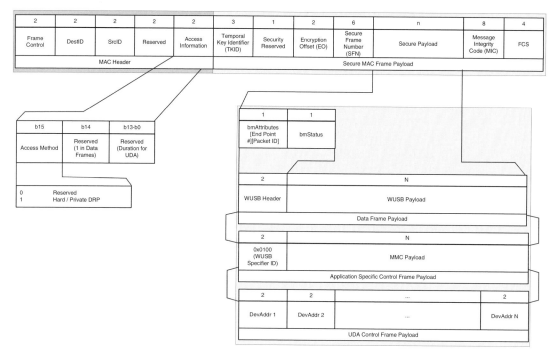

Figure 9.9 WUSB frame structure.

dependent parameters, as shown in Figure 9.10. WUSB uses only a portion of the frame control structure and augments this with additional control structures within its payload fields.

The most important fields, functionally, are the Frame Type, Frame Subtype, which is also overloaded as Delivery ID, and Acknowledgement Policy.

WUSB only uses data- and application-specific control frames. Delivery ID is always set to the stream index chosen for this WUSB channel by the WUSB host. WUSB has its own acknowledgement and retransmission mechanism, and so ACK Policy is always set to NO-ACK.

9.2.4.2 Secure Frames

The basic frame structure of the MAC allows a payload to be transported transparently by the PHY from one device to another. No assumption is made about the content of the payload other than whether it is nonsecure or secure. If the Secure bit is set in the frame control field of the MAC header, the payload is in secure frame format and is formatted as shown in Figure 9.11.

The secure frame header carries encryption key and secure message integrity check information, as well as a secure frame counter to protect against replay attacks. A secure frame control field allows the transmitting device to send part (or all) of the

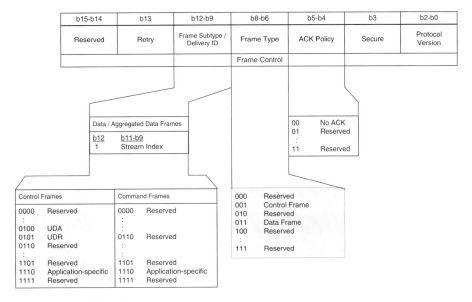

Figure 9.10 Frame control fields.

secure frame header and secure payload unencrypted. Encryption offset (EO) defines the starting position in the frame payload where encryption begins.

9.2.5 Host Role

The WUSB host differs from its wired counterpart in several fundamental ways. The first concerns the definition and management of the communications channel, and several others result from the unreliable nature of this channel.

9.2.5.1 WUSB Channel

The WUSB channel defines the communications channel between the WUSB host and its devices and consists of a sequence of active transaction periods, each identified by an initial control frame called a microscheduled management command (MMC) (see Section 9.2.5.2). Unlike most communications channels, the WUSB channel is continuous in time, but the active periods are not necessarily contiguous.

3	1	2	6	n	8
Temporal Key Identifier (TKID)	Security Reserved	Encryption Offset (EO)	Secure Frame Number (SFN)	Secure Payload	Message Integrity Code (MIC)
Secure MAC Frame Payload					

Figure 9.11 Secure frame format.

262 Chapter 9

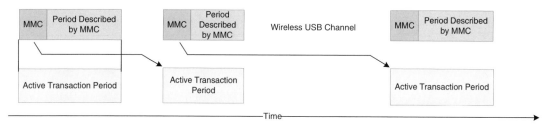

Figure 9.12 WUSB channel.

Each control frame describes the operations to be performed and then points to the location of the next control frame. All references are relative to the beginning of the control frame. (See Figure 9.12.)

9.2.5.2 Microscheduled Management Command

The control frame at the beginning of each active period on the WUSB channel is called a microscheduled management command, or MMC. The MMC is a secure frame with encryption offset (EO) set to the length of the payload. Hence even unconnected devices may receive and decode MMCs. As Figure 9.13 shows, the MMC consists of a header, a sequence of information elements (IEs), and a link to the next MMC. The sequence of IEs describes what takes place in the interval immediately following the MMC. References to MMCs in the following text should

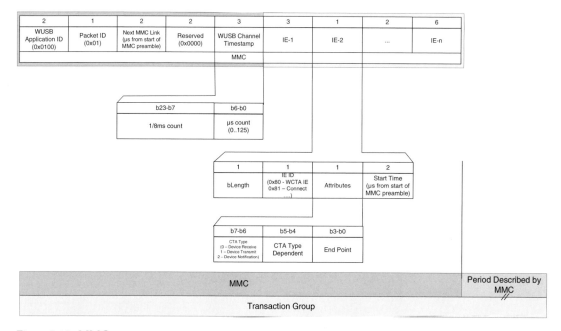

Figure 9.13 MMC structure.

be taken to mean the MMC data structure itself or the MMC and the interval following it as described by the IEs.

The chain of MMCs constitutes the WUSB channel.

The MMC contains wireless channel time allocations (WCTAs), a name taken from the 802.15 specification for reserved channel time. For each WCTA IE in the MMC there is a protocol time slot allocated for the exchange of data described by the IE. There are three main types of IE, device receive, device transmit, and device notification. The IE defines the start time of the transfer so that a WUSB device addressed by the WCTA knows exactly when it needs to be ready either to receive or to transmit data. The duration of the protocol time slot is implicitly defined by the start time of the following protocol time slot.

9.2.5.3 Transaction Group

Since changing from receive to transmit in a high-speed radio subsystem takes a nontrivial amount of time, the transaction group is ordered to minimize the number of such changes. The host transmits data to the cluster in the first part of the transaction, cluster member devices transmit to the host in the second part of the transaction group, and, finally, a handshake phase permits devices to acknowledge data received from the host. Note that each of these phases of the transaction group may be empty.

To permit unassociated devices to request membership in the cluster and to support reassociation of devices that have been idle, the host includes an appropriate number of device notification time slots (DNTSs). The host schedules these randomly among the transaction group time slots to mitigate against hidden interference cases in which a device may not transmit in an announced notification time slot as it would interfere with one of its neighbors.

9.2.5.4 Data Flow

Each WUSB device informs the WUSB host of its maximum input/output capabilities during the association procedure. Data is transferred using the PHY burst mode, a special mode where the PHY provides tight timing between successive frames transmitted by the same device, and the WUSB device declares the size and maximum number of frames it can handle in a data burst. (See Figure 9.14.)

Timing of inbound data is protected by increased guard bands within the protocol time slot since each device measures intervals using its own local clock, which is not synchronized with the host's clock. However, time is reset by each device at the start of the MMC; hence, the guard times are kept as small as possible to increase efficiency.

WUSB provides acknowledgement of data-frame exchange within the MMC structure. Each WCTA acknowledges preceding data frames transmitted to the host. Data frames transmitted to devices are acknowledged by explicit handshake transmissions at the end of the transaction group.

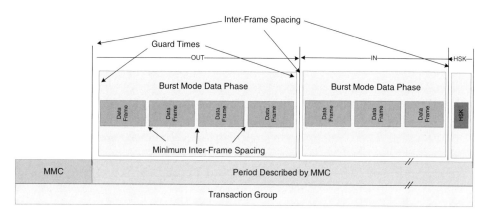

Figure 9.14 WUSB data phase.

Unacknowledged frames are retransmitted unless their validity is deemed to have been exceeded. The WUSB specification includes support for isochronous data streams by providing both presentation time and a sliding window with a handshake mechanism.

9.2.6 Device Role

9.2.6.1 Host Association

Before the device can become active, it must associate with a host by using one of the host's declared DNTSs. Access to the medium during a DNTS is via slotted Aloha protocol. The device uses the predefined unassociated device address in the transmitted association request frame.

The device informs the host of a number of parameters, including whether it includes a WiMedia system component and is capable of managing its own interference environment. Interference is managed by the WiMedia MAC Beacon and Distributed Reservation protocols. (See Sections 9.3.3 and 9.3.4.)

If the device has no WiMedia system component, the interference environment must be managed by the host. Since there is no guarantee that the host's and device's neighborhoods are the same, the host can instruct the device to transmit and receive beacons.

9.2.6.2 Directed Transmission and Reception

The host maintains a benign interference environment via its WiMedia MAC system component. This component executes the WiMedia Beacon Protocol to signal its presence and declare reservations covering the time it uses for its MMCs and associated transaction groups. The reservations are established and maintained by the WiMedia Distributed Reservation Protocol (DRP).

The Beacon Protocol requires each active device (a device is active if it transmits within the WiMedia superframe) to send a beacon frame in a selected beacon slot at the start of the superframe. The Beacon Protocol resolves beacon slot conflicts so that each active device has a separate beacon slot. In addition, the Beacon Protocol requires that devices listen for beacons and interpret their content to be able to respect declared reservations and to maintain information about the neighborhood.

The directed transmission and reception feature of WUSB enables a WUSB host to provide a preformed frame for transmission or to have the device listen for a frame at an instant defined by the host. Since the host WiMedia system component maintains the beacon period timing, the WUSB host can construct a beacon frame for transmission by a nonbeaconing device and receive beacon frames via directed reception.

9.2.7 System Component

WUSB devices may, and WUSB hosts must, implement a system component providing support for the WiMedia Beacon and Distributed Reservation protocols in order to provide a benign interference environment for both WUSB and mixed WUSB and WiMedia systems.

The WUSB channel is managed by the WUSB host by arranging the start time of each MMC to be within a reserved time owned by the host and maintained by its system component. The transaction group following the MMC must also fall entirely within a block of reserved time.

WiMedia devices maintain a different channel structure, using a repeating fixed-length superframe instead of a monotonically increasing channel time.

The two channels, WUSB and WiMedia, can coexist in the same RF spectrum at the same time and within range provided that the chain of WUSB MMCs and their associated transaction groups fall entirely within blocks of time reserved by the host using the WiMedia DRP. (See Figure 9.15.)

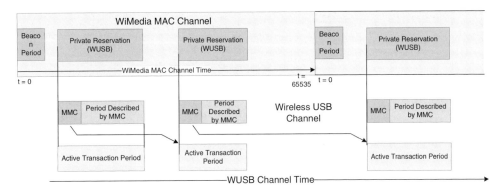

Figure 9.15 Channel sharing.

The WUSB host system component maintains alignment of the start of the WiMedia MAC superframe according to a slowest-neighbor rule (see Section 9.3.3.1). Since any correction must be applied within each superframe according to the WiMedia MAC Beacon Protocol, the WUSB host must coordinate the construction of the MMC forward pointer such that it takes into account any correction for superframe end times.

9.3 WiMedia MAC

Whereas WUSB has its origins in USB and provides cable-free operation for a set of associated devices, the WiMedia MAC has its origins in consumer electronics equipment that needs to operate in a more dynamic environment, including support for multiple collocated, but independent, networks, some of which may be mobile. The WiMedia MAC is published by the WiMedia organization (www.wimedia.org).

These needs lead to a fully distributed system design based on a repeating, fixed-length superframe. As shown in Figure 9.16, the superframe begins with a beacon period, and the remainder of the superframe is used for data transfer. All active devices transmit beacon frames during the beacon period. The Beacon Protocol coordinates device operation and supports device discovery, mobility functions, and management of the dynamic communications environment.

The major difference between WUSB and the WiMedia MAC is that whereas the WUSB system design focuses on the WUSB host and its set of associated devices, which form the WUSB cluster, the WiMedia MAC only considers a device and its neighbors. According to the WUSB architecture, either the host role or the device role is executed in the WUSB channel. According to the WiMedia MAC architecture, every device is identical and executes all required protocol functions.

The symmetry of device operation is captured in two concepts:

- The beacon group consists of a device and its neighbor devices with aligned beacon period start times (BPSTs).
- The extended beacon group consists of a device and its neighbors and their beacon groups.

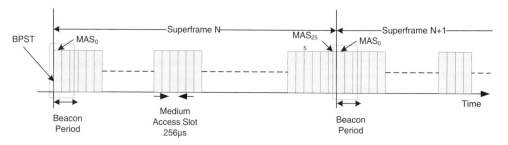

Figure 9.16 WiMedia MAC superframe structure.

When a WiMedia MAC device begins operation, it searches one or more RF channels for beacon frames. After selecting an RF channel for its operation, if there are existing devices within range, their beacon frames define the start of their superframe, as well as describing the devices and their neighborhood information. The new device aligns its superframe operation to that of the existing neighbor(s). If no existing devices are detected, the new device begins transmitting its beacons and defines its own superframe start time.

Figure 9.17 shows a decentralized device arrangement. Device A has a beacon group consisting of Devices B, C, and D. Its extended beacon group includes Device E but not F. The occupancy of the beacon period will be different for these devices, as we shall see in Section 9.3.3.

The WiMedia MAC protocols ensure that only one device in the extended beacon group transmits a beacon frame at a time. The control information carried in the beacon frames provides for spatial reuse of the communications channel by ensuring that a device does not transmit while any of its neighbors are transmitting or receiving. Even if communication is asymmetric, the Beacon Protocol ensures correct neighborhood management.

Information carried in the beacon frames includes device parameters, reservation declarations, neighborhood information and a variety of requests and responses expressed as IEs.

The WiMedia MAC defines three medium access protocols:

- Beacon access, which is exclusively reserved for use during the beacon period
- DRP access, where specific devices have priority access to the medium

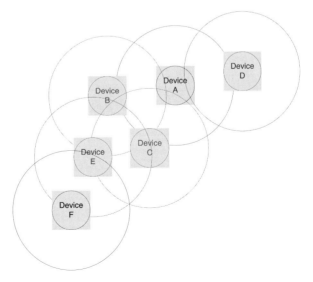

Figure 9.17 Decentralized device arrangement.

- Prioritized contention access (PCA), which is a random-access mechanism based on 802.11e

To support common references to time resources, the superframe is broken down into a fixed number (256) of fixed-length (256 μs) medium access slots (MASs) for a total superframe length of 65,536 μs, or approximately 65 ms.

9.3.1 WiMedia MAC Frame Structure

The WiMedia MAC data structures are carried in WiMedia PHY PDUs, as shown in Figure 9.18. Timing information is maintained by reference to the PHY PDU preamble, and the payload length is carried in the PHY header. The PHY header protection covers the PHY and MAC header fields.

The payload can carry up to 4,095 unencrypted data octets or 4,075 encrypted data octets and a 20 octet secure header and message integrity code. Both payload forms are followed by a four octet frame check sequence.

The MAC header provides coding for a number of frame and subframe types, source and destination addresses, and sequence control for both normal and fragmented frames. Since a device may receive unexpected frames (for example, owing to mobility effects) Access Information includes an indication of the access method.

9.3.1.1 WiMedia MAC Frame Control

There are six main frame types (see Figure 9.19):

- Beacon frames, which are only used in the beacon period to support the Beacon Protocol
- Control frames, which are used to support frame transactions and release unwanted reserved time
- Command frames, which may be used to establish reservations, probe the capabilities of neighbor devices, exchange security information, and perform range estimation between pairs of devices

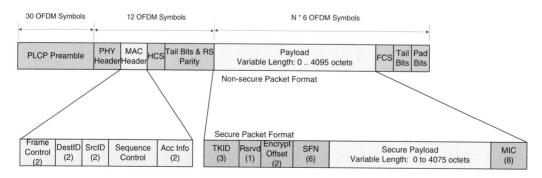

Figure 9.18 MAC frame format.

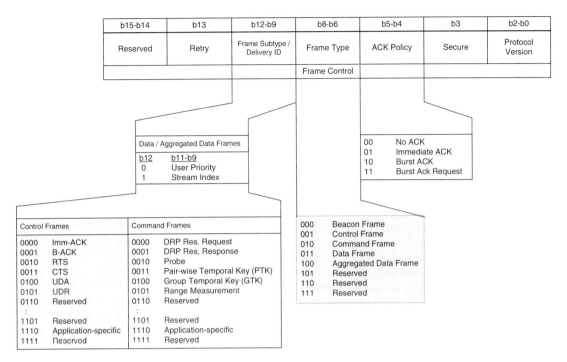

Figure 9.19 Frame control fields.

- Data frames
- Aggregated frames, where multiple MAC Service Data Units (MSDUs) to the same destination may be carried in a single payload
- Application specific frames

Because the radio medium is unreliable, all frames are numbered to allow recipient devices to detect missed or duplicated frames. If the radio link between the transmitter and receiver is poor, a frame may be fragmented into smaller units to increase transfer success. Each fragment is sent in its own frame but with the same sequence number. The sequence control structure of frame control to support these facilities is shown in Figure 9.20.

Access Method (Figure 9.21) includes coding of whether the frame was transmitted in a reservation of type Hard or Private or transmitted by the owner of a Soft reservation when accessing the medium without back-off (see Section 9.3.4.1).

b15	b14	b13-b3	b2-b0
Reserved	More Fragments	Sequence Number	Fragment Number

Figure 9.20 Sequence control field.

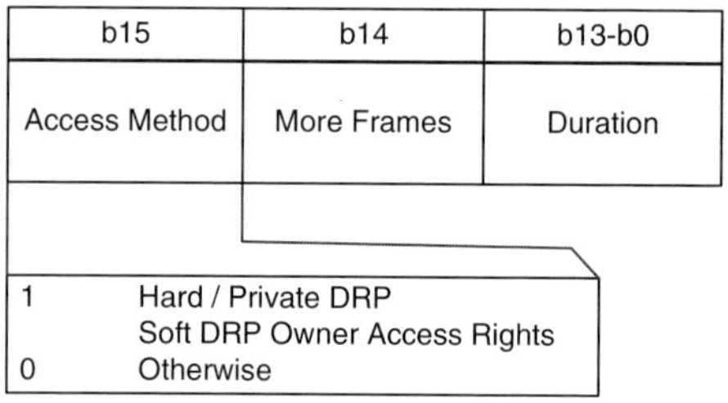

Figure 9.21 Access information fields.

More Frames is used to assist in power saving. A device will clear this control bit if it will not send further frames to its target in a reservation block or will not attempt to send further frames to the same recipient using the PCA method in this superframe.

Duration is also used to support power saving and indicates the expected interval during which the medium will be busy for this frame transaction. Devices not addressed by the frame transaction may enter power-saving states for this interval.

9.3.1.2 Secure Frames

If the Secure Payload bit is clear, the frame payload contains the MSDU (or a fragment). If this bit is set, the frame payload is protected using the Advanced Encryption Standard (AES) algorithm for both authentication (via computation of the message integrity code) and encryption of the secure payload itself.

Session keys can be securely generated and exchanged from master keys, which are distributed by some out-of-band means.

An interesting feature of the secure payload is the EO, which identifies the start octet within the secure payload where encryption will begin. This allows some or all of the secure payload to be transmitted unencrypted while maintaining authentication of the data via the Message Integrity Code (MIC).

9.3.1.3 Acknowledgment Policies

There are three acknowledgement policies for WiMedia MAC frame transmissions:
- No acknowledgement
- Immediate acknowledgement
- Block acknowledgement

Immediate acknowledgement requires the recipient device to present an ACK frame within a short interframe space (SIFS) interval (10 μs) after the end of the transmitted frame. Absence of this ACK frame causes the transmitter to discard or retransmit the frame based on system-determined policies.

The block acknowledgement (B-ACK) is a powerful optimization where an initial exchange of a frame with a B-ACK request set in the frame control field allows a recipient to inform the transmitter how much buffer space it has available and how many frames it is prepared to receive. The transmitter may then send up to that number of frames without any constraint on when or in which superframe, limited only by the total number of frames the recipient device has indicated that it is willing to receive. The sequence is terminated when the transmitter includes the B-ACK request acknowledge policy, and the recipient sends a B-ACK response control frame with a map of which frames have been successfully received. The B-ACK response may again indicate the number of frames and total buffer space available at the recipient, and data exchange may continue.

It is important to note that the recipient device has control and can always limit the number of frames and total data octet count that a transmitter may send to it.

9.3.1.4 Fragmentation, Reassembly, and Aggregation

Because the link between two devices will change dynamically, a transmitting device may optimize a number of parameters, including payload size, data rate, and transmitted power, to maximize the likelihood of a successful frame exchange.

An MSDU accepted for transmission by the MAC may be fragmented into up to eight pieces of arbitrary size and sent as a sequence to the destination device. The recipient must successfully receive all of the fragments and reconstruct the original MSDU before delivering it to the MAC client. If any fragment is lost, the entire MSDU transmission is lost.

Similarly, if a link is strong, the transmitter may aggregate multiple MSDUs to the same destination and transmit them in a single frame, thereby saving interframe time and individual acknowledgements for greater protocol efficiency.

9.3.1.5 Device Addressing

Device identifiers are large and, if used as addresses, would be far larger than the maximum number of neighbors that can be supported by the WiMedia MAC. Instead, the WiMedia MAC uses 16 bit dynamically allocated addresses, or DevAddrs.

The 16 bit address space is divided into four regions (see Table 9.2).

A device chooses its DevAddr such that it is unique in the device's two-hop neighborhood. Addresses are selected at random using a uniform distribution over the

Address Value	Description
0x0000..0x00FF	Private Address Range
0x0100..0xFEFF	Generated Address Range
0xFF00..0xFFFE	Multicast Address Range (McstAddr)
0xFFFF	Broadcast address (BcstAddr)

Table 9.2 WiMedia MAC Address Ranges

range and verified against being in use by reference to received beacons and the beacon period occupancy IEs (BPOIEs) they carry. (See Section 9.3.3.)

Devices not using the WiMedia protocol but WiMedia frame formats can use addresses from the Private address range. Normal devices use the Generated address range. The McstAddr range is used for multicast addresses. Before a frame is transmitted using a multicast destination address, a binding must be declared between a multicast EUI-48 and a McstAddr. This enables recipient devices to establish filter tables for frames destined to EUI-48 multicast groups of interest to their clients.

There is one reserved address, 0xFFFF, the broadcast address.

9.3.2 Data Transfer

Data is exchanged between devices in frame transactions as shown in Figure 9.22. A frame transaction may begin with a Request to Send/Clear to Send (RTS/CTS) exchange, which, although expensive as it is sent at the lowest payload rate, improves transmission success in environments susceptible to interference from hidden terminals. RTS/CTS can be useful in a highly mobile environment, for example.

Between each frame where there is a change of transmitter (i.e., between the RTS and CTS control frames, for example), the medium must be free for at least, and in certain cases exactly, a SIFS interval, defined as 10 μs. This is sufficient time for the transmitter to switch its radio into receive state and vice versa.

Following the RTS/CTS is one frame, possibly with immediate or B-ACK acknowledgement, or several frames with associated B-ACK acknowledgment policy as described above.

If several frames are transmitted from the same source, the interframe space may be shortened to the minimum interframe space (MIFS) with the assistance of the PHY to ensure accurate symbol timing of exactly six symbols (1.875 μs). The combination of MIFS spacing and B-ACK acknowledgment allows significant improvement in protocol efficiency provided the frame error rate is low.

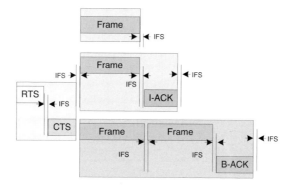

Figure 9.22 Frame transaction.

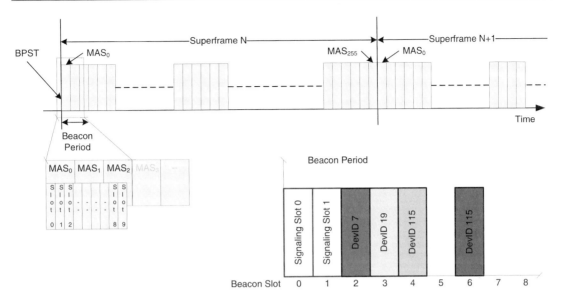

Figure 9.23 Beacon period and beacon slot timing.

9.3.3 Beacon Protocol

The superframe is defined to begin at the start of the first beacon slot at the instant known as the BPST. Beacon slots are slightly shorter than one-third of a MAS, being 85 μs long. As Figure 9.23 shows, beacon slots are contiguous, giving a small unused time in the last MAS of the beacon period.

In normal operation, up to 16 MAS are allowed to be used for beacons. The actual number used is selected dynamically, and beacons are compacted so that they always occupy the minimum number of slots necessary. A beacon is deemed no longer to be valid if it is missing after three successive superframes.

The Beacon Protocol is the main communications management protocol in the WiMedia MAC. It is used for device discovery, as a broadcast control channel, and as an aging mechanism.

Beacons are sent in their own frame type (type 0) and contain a beacon parameter block and a BPOIE and zero or more other IEs (Figure 9.24). The beacon parameter block carries the device's unique device identifier (a EUI-48) and the beacon slot

8	L_1	L_m	...	L_n	4
Beacon Parameters	Information Element-1	Information Element-2	...	Information Element-n	FCS
Beacon Frame Body					

Figure 9.24 Beacon frame payload.

Octets: 1	1	1	K	2	...	2
Element ID	Length	BP Length	Beacon Slot Bitmap	DevAddr 1	...	DevAddr n
Beacon Period Occupancy IE							

Figure 9.25 Beacon period occupancy IE.

number the device is using to send its beacon (see Superframe Synchronization later in this chapter for an important use of the beacon slot number). It also includes additional control information to signal other beacon-management functions and the device security mode.

The most important IE for neighborhood management is the BPOIE. This IE carries the local device view of its neighborhood and allows each device to build its view of the two-hop neighborhood and, hence, to regulate access to the medium for the reservation access mechanism.

The format of the BPOIE is shown in Figure 9.25 and includes the local device's declared beacon period length, a bitmap of occupied beacon slots, and the DevAddrs of the transmitter of each beacon received. Including the DevAddr in this structure allows the dynamically assigned device address mechanism to resolve duplicate addresses in the two-hop range and to maintain stable communications.

Figure 9.26 shows how the beacon period in the example distributed arrangement in Figure 9.17 will be reported in the BPOIE of each device. Note that each device has its own local view of the network topology, that the views are not the same, and that two devices may use the same beacon slot if neither is in the extended beacon group of the other.

A device may also send its beacon in one of the signaling slots. Since all devices must listen at least up to their own beacon slot, all devices listen to the signaling slots. A new device, possibly using a beacon slot higher than any neighbor is listen-

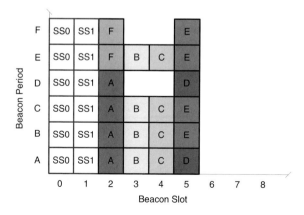

Figure 9.26 Device-centric neighborhood.

ing to, will be detected via its beacon in the signaling slot, and that beacon will declare the real slot being used by the new device.

A similar mechanism is used when two devices detect that they are using the same beacon slot, in which case they select new slots and redeclare them. Periodically, each device listens to its own beacon slot to see if it is in collision in an otherwise undetectable situation.

9.3.3.1 Superframe Synchronization

Since devices always align the start of their superframes to those of their neighbors, the inclusion of the beacon slot number in the beacon frame allows the recipient device to calculate whether the transmitter's clock is faster or slower than its own. A beacon that arrives earlier than the local nominal beacon slot start time is sent by a device with a faster clock, and one that arrives after the start is sent from a device with a slower clock. This is important as the slowest device in the neighborhood determines when the superframe ends (since it will have the longest superframe in absolute time). All neighbors of the slowest device extend their superframes so that the BPST alignment is maintained for all devices. (See Figure 9.27.)

9.3.3.2 Beacon Period Merging

Since all devices must align their BPSTs to those of their neighbors, there can be only one beacon period. The size of the data that the beacon must carry, together with the time it takes to transmit it in the fixed-size beacon slot, imposes a limit of 96 devices per beacon period, although there is a normal operation policy that imposes a limit of 48 devices.

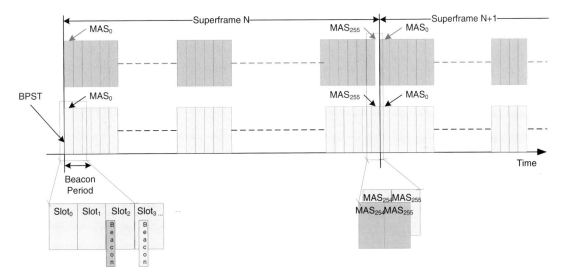

Figure 9.27 Superframe alignment.

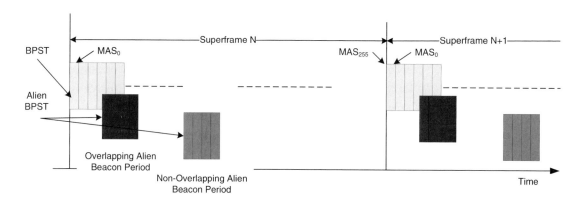

Figure 9.28 Beacon period merging.

Owing to mobility and other propagation effects, previously out-of-range devices may come into range of each other. In this case, the alien BPSTs are likely to be unaligned and must be merged. (See Figure 9.28.)

A beacon period merging protocol is defined where a device, receiving a beacon frame with a BPST unaligned with its own (an alien beacon), first determines whether the two beacon periods overlap. If they do, the device immediately moves its beacon to a position equivalent to concatenating the two beacon periods to avoid destroying the critical management information sent in beacon frames. (See Figure 9.29.)

If the beacon periods are unaligned but do not overlap, the device receiving the alien beacon determines whether it should initiate moving its BPST to the alien beacon period depending on whether its BPST falls in the first half of the alien superframe.

A protocol is then initiated where the device includes a beacon period switch IE in its beacon, indicating the new BPST and where in the new beacon period it will

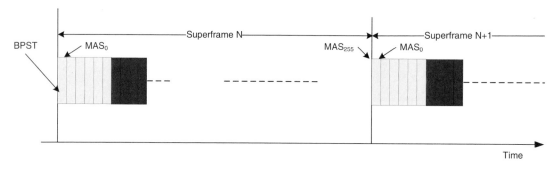

Figure 9.29 Beacon relocation.

send its beacon, together with a countdown indicator of the superframe in which the switch will occur. If new or updated alien beacon information is received, the switch protocol is restarted. Devices receiving beacon period switch IEs from neighbors similarly initiate a merge operation and provide a coordinated relocation of the beacons in the two beacon periods. Although more complex than an immediate beacon relocation this merge protocol may reduce disruption to established traffic flows.

9.3.4 Distributed Reservation Protocol

The DRP exploits the power of the Beacon Protocol to negotiate conflict-free access to the medium in the distributed WiMedia network. All devices must respect reservations declared by their neighbors and avoid interfering with their communications.

DRP reservations are negotiated by including DRP IEs in beacon frames. A request/response protocol allows a device (the reservation owner) to suggest specific MASs to be reserved for a data flow identified by a stream ID. The target of the reservation request may be a single DevAddr or a McstAddr. The number of MASs reserved determines the capacity of the reservation, and the distance between reservation blocks determines the latency of medium access. Devices can attempt to reserve MASs that match the traffic profile of their applications. (See Figure 9.30.)

A target device of a reservation request may be unable to accept the offered reservation owing to an existing active neighbor reservation or for another reason. The request in this case is refused by responding with an appropriate reason code and including a DRP availability bitmap describing the MAS the target device may accept for a new reservation request. If the target device responds by including an identical DRP IE to that requested, then the reservation is established.

DRP IE declarations use a zone representation of the MAS of the superframe. In this view, the 256 MASs are organized into 16 zones starting from the BPST, where each zone consists of 16 consecutive MASs. Zone 0 consists of MASs 0 to 15, zone 1 contains MASs 16 to 31, and so on. (See Figure 9.31.)

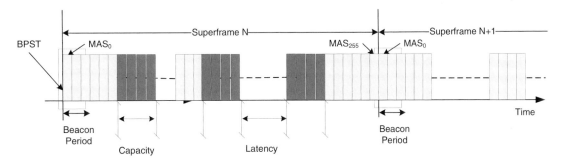

Figure 9.30 Reservation capacity and latency.

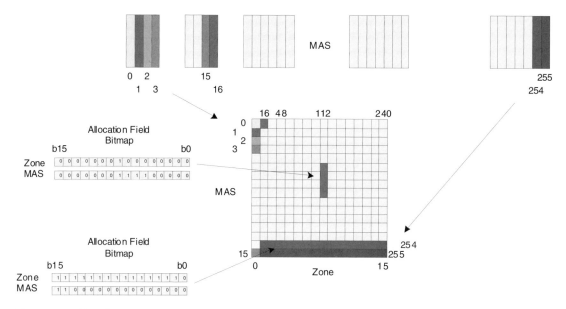

Figure 9.31 Zone bitmap representation.

A DRP IE declares a reservation with a 16 bit zone bitmap and a 16 bit MAS bitmap. For example, a reservation with MAS 14 and 15 in each zone from 1 to 15 would be coded as (0xFFFE, 0xC000), and a single reservation of four consecutive MAS numbers 117 to 120 would be coded as (0x0100, 0x01E0).

In selecting the MAS to use for the reservation, a device takes into account the application's service interval, which will be transformed into the number of MASs between each reservation block, and the bandwidth required for data transfer, which will correspond to the number of MASs in each reservation block. A reservation with MAS reserved in each consecutive zone will correspond to a service interval of 4 ms (16×256 µs).

9.3.4.1 Reservation Types

A number of different types of reservation can be established to satisfy the various kinds of data communications between devices. In each case, one or more MASs is reserved in a reservation block and declared in a DRP IE in the device's beacon.

Hard Reservation
Communication within a Hard reservation is well defined and quite restrictive. Only the reservation owner is permitted to initiate frame transactions within the reservation boundaries. Time that would otherwise be unused by the reservation owner or used for acknowledgements may be released for general contention access by transmitting a special Unused DRP Reservation Announcement control frame,

which is followed by an Unused DPR Reservation Response frame from each of the Hard reservation's targets.

Soft Reservation

A Soft reservation is more flexible in transmission permissions. The Soft reservation owner has priority access to the medium during the Soft reservation, but a reservation target may also transmit within the reservation if the owner allows the necessary interframe space to be idle.

Access to the medium follows the rules of the Prioritized Contention Access Protocol, where devices wait for an idle period determined by the priority of the data to be sent. Higher-priority data with shorter idle periods gains access to the medium earlier than lower-priority data with correspondingly longer idle time requirements.

Private Reservation

A Private reservation may be established by any device, and within the reservation, any medium access rules may be applied. This is the principle mechanism for spectrum sharing provided by the WiMedia MAC. WUSB establishes its channel in Private reservations.

PCA Reservation

A PCA reservation is special in that MASs declared within the reservation may be declared by multiple owners. In other words, there is no reservation collision for PCA reservations.

The main reason for this reservation type is to support two-hop contention resolution within declared MASs, which may be used by any device in the beacon group. Permitting collisions in the DRP IE means that a set of MASs can be reserved for PCA access, and even if an owner moves out of range, a similar reservation maintains the MAS availability for PCA.

Alien Beacon Reservation

Finally, a special reservation type is reserved for protecting alien beacon periods during the beacon period merging procedure. Alien beacon reservations have priority over all other reservations and require no reservation negotiation.

9.3.5 Contention Access Protocol

The final medium access mechanism in the WiMedia MAC is the PCA Protocol, which allows a device to contend with other devices for access to the medium during PCA reservations or in any unreserved MAS. PCA is based on the IEEE 802.11e Enhanced Distributed Channel Access mechanism with four access categories and contention windows computed for expected WiMedia topologies.

A device using PCA to attempt transmission of a frame should include a traffic information map IE in its beacon, informing potential recipient devices that they should listen for PCA traffic in MASs they have identified as available for PCA traffic.

9.3.5.1 Collision Avoidance

PCA operates by sensing the medium before attempting to transmit a frame and performing a back-off procedure to randomize the delay with which a subsequent access attempt will be made. Figure 9.32 shows the relationship between different traffic priorities and the back-off algorithm intervals.

In addition to the SIFS and MIFS, PCA adds four arbitration interframe spaces (AIFS). The shorter the duration of the AIFS, the higher the traffic priority. The four access categories are Voice (AC_VO), Video (AC_VI), Best Effort (AC_BE), and Background (AC_BK). When a frame is placed in back-off, a number of back-off slots is chosen at random with linear probability from a range dependent on the traffic priority. Each time the medium is idle for at least AIFS[i], the ith priority traffic can begin counting down the back-off slots. When all of the back-off has passed, the device has gained a transmit opportunity and can attempt to send its frame.

To avoid unnecessary listening to the medium, the duration field of Access Information is used to maintain a virtual busy network allocation vector (NAV). As long as the NAV is greater than zero, the device is forbidden to transmit.

9.3.5.2 Internal Collisions

In PCA, the multiple internal queues (one for each priority) contend for access to the channel with different interframe spacing (from which the queue priority is

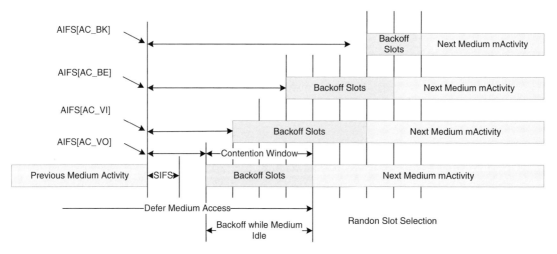

Figure 9.32 PCA medium access categories.

determined) and with independent back-off windows (meaning independent random wait intervals).

Since the values of the back-off are selected at random and the unit of countdown is quantized, it is possible for two (or more) frames to arrive at the head of their back-off queues at the same unit of time. This is called an internal collision.

These collisions are processed in the same manner as real physical collisions would be processed, except for the actual transmission of the lower-priority frames.

The frame from the highest-priority queue always wins an internal collision.

9.3.6 Ranging

UWB waveforms posses useful time-domain properties that can be exploited to derive distance estimation.

A ranging reference point is defined in the WiMedia PHY—the first sample of the first symbol of the channel estimation sequence of the frame preamble in both the transmit and receive directions.

The PHY captures a high-speed counter value when this reference point is detected, and this counter value is made available to the MAC.

The MAC supports the exchange of these ranging counter values by the exchange of specific ranging measurement command frames.

The range measurement exploits the known timing relationship between a frame and its immediate acknowledgement to capture one or more range measurement pairs.

The timing counter values at the ranging initiator and responder can be used to derive the distance, knowing the speed of radio frequency emissions.

9.3.7 Power Saving

The principle mechanism for power saving is to enter a low-power or hibernation mode when scheduled communication is known to be absent. Hibernation, supported by the hibernation IE and hibernation anchor IE, allows a device not to transmit any beacon frames for a specified number of superframes while still maintaining its place in the beacon group. The hibernation anchor effectively acts as a surrogate for the hibernating device by including that device's beacon information even though it has not received the beacon from the hibernating device.

The reservation protocol and various control bits indicating more frames also assist devices in scheduling power-saving modes.

9.3.8 Transmit Power Control

Although the levels permitted for UWB emissions are very low, there is some scope for transmitter power control (TPC).

The WiMedia MAC specifies that beacon frames, which are used to carry the most important signaling information for maintaining channel access stability, should be transmitted at the highest power and the lowest data rate, thus reaching the maximum range.

All other frames should, in general, be transmitted at the minimum power necessary to support the link. The link management IE may be used to inform a device of the quality of the link between a pair of devices. Although the link is not guaranteed to be stable for any length of time, statistical use of the link management IE, together with adaptive data rate, can be used to optimize communications between devices.

9.4 Conclusion

In this chapter, we have identified key features of UWB systems that impact the design of the medium access and control protocols for UWB systems. We have examined two classes of systems, centralized and distributed, and described the key features of the two current MAC specifications for UWB systems, the Certified WUSB specification and the WiMedia Distributed MAC specification.

The WUSB design places full control over medium access scheduling in the WUSB host system, while the WiMedia MAC specifies a fully distributed architecture with advanced mechanisms to support contention-free reservation of access to the medium without central coordination.

Use of a common beacon protocol enables the radio resource to be shared among multiple protocols, effectively acting as a common signaling protocol.

Both architectures exploit the high instantaneous data rates offered by UWB transmission techniques to provide high-quality-of-service communications services to their clients even though each design is fundamentally different in its approach to medium access management.

Bibliography

Distributed Medium Access Control (MAC) for Wireless Networks, Release 1.0, WiMedia Alliance, Inc., December 2005.

IEEE 802.15.3-2003, September 2003, IEEE Standard for Information Technology—Telecommunications and Information Exchange between Systems—Local and Metropolitan Area Networks—Specific Requirements, Part 15.3: Wireless Medium Access Control (MAC) and Physical Layer (PHY) Specifications for High Rate Wireless Personal Area Networks (WPANs).

IEEE 802-2001, March 2002, IEEE Standard for Local and Metropolitan Area Networks: Overview and Architecture.

Multiband OFDM Physical Layer Specification, Release 1.1, July 2005, WiMedia Alliance, Inc.

Universal Serial Bus Revision 2.0 Specification, April 2000, USB Implementers Forum, Inc., at www.usb.org/developers/docs/usb_20_02212005.zip.

Wireless Universal Serial Bus Specification, Release 1.0, May 2005, USB Implementers Forum, Inc., at www.usb.org/developers/wusb/docs/WUSBSpec_rc10.pdf.

10

Standards for UWB Communications
by Jason L. Ellis

There are several types of industry standards, namely, open standards, which are solutions developed through collaborative industry efforts from the ground up; proprietary standards, which are specifications that have been developed through private efforts and heavily promoted to an industry forum (typically without modification); and de facto standards, which result from wide industry adoption of a proprietary solution not endorsed by an industry forum. Examples of each follow, respectively. WiFi technologies, whose development started with 29 attendees from numerous companies at the inaugural Institute of Electrical and Electronics Engineers (IEEE) 802.11 meeting in Ontario, Canada, in September 1990 [1]; Code Division Multiple Access (CDMA) technology promoted by Qualcomm Corporation and published as IS-95 in 1993; and Dolby (A-type and B-type) noise reduction offered by Dolby Laboratories as early as 1966 [2].

This chapter explains why standards are important for consumers and technology integrators, details open standards and proprietary standards, and provides an overview of the various industry forums, all the while explaining the development of UWB-based communications standards.

10.1 Importance of Standards

It is recognized that for more cases than not, industry standards are required to enable mass-market opportunities in the consumer segment. Standards allow for multiple-vendor interoperability, which creates competitive products that afford end consumers with choice. Such choice targets consumer priorities, including price, performance, and brand recognition. For original equipment manufacturers (OEMs) and original design manufactures (ODMs), standards reduce risk by providing multiple sources for technology and enable markets more quickly due to the development of competitive products. The IEEE defines a standard as

> "a published document that sets out specifications and procedures designed to ensure that a material, product, method, or service meets its purpose and consistently performs to its intended use. Standards solve issues ranging from product compatibility to addressing consumer safety and health concerns. Standards also simplify product development and reduce

non-value-adding costs thereby increasing a user's ability to compare competing products. They also are fundamental building blocks for international trade. Only through the use of standards can the requirements of interconnectivity and interoperability be assured and the credibility of new products and new markets verified enabling the rapid implementation of technology." [3]

The International Organization for Standardization (ISO) remarks that "if there were no standards, we would soon notice. Standards make an enormous contribution to most aspects of our lives—although very often, that contribution is invisible. It is when there is an absence of standards that their importance is brought home. For example, as purchasers or users of products, we soon notice when they turn out to be of poor quality, do not fit, are incompatible with equipment we already have, are unreliable or dangerous. When products meet our expectations, we tend to take this for granted. We are usually unaware of the role played by standards in raising levels of quality, safety, reliability, efficiency and interchangeability—as well as in providing such benefits at an economical cost. [4]

As previously mentioned, standards are created as a result of the promotion of proprietary technology or from defining technology from the ground up in order to best meet the market requirements; there are trade-offs supporting each approach with regard to time to market, innovation, and technical due diligence.

10.2 Standards Development

Companies sometimes develop technology having strictly concluded only technical merit, without confirming a true market need. Sometimes a market need arises, and it is only then that technical innovation is encouraged. However, the process of introducing new technology features as a means to adding new user benefits is actually quite iterative, and industry forums provide the venue for such discussion and iteration. Within industry forums, the right mix of participation requires involvement from technological innovators, solution providers, and consumer product companies. It is there that technological capabilities can be shared such that new usage models and applications can be conceived, resulting in a set of application and technical requirements that yield products with added value. To be true to the process, technology innovators must be prepared to make changes to the technology to reflect customer requirements. Therein lies the challenge as sometimes the need to commercialize technology and to maintain time to market and competitive advantage takes priority over iterating or completely redesigning the solution.

This predicament is ever-present in the commercialization of UWB solutions, especially for single-product companies that have no other means of revenue and are financed by venture capital. If those companies slow down their development to accommodate new requirements set before them by OEM/ODM customers, these startup companies fear going out of business from lack of revenue or disinterested financers. If they choose not to incorporate those requirements in their first-generation products, and if they fail to deliver their solutions in a timely manner, new competitors may be able to introduce far superior approaches that gain momentum. Likewise, if a market is not mature enough to utilize early solutions, the pioneering companies may also go out of business. This story is definitely that of timing, not being too early to market, not being too late with product, and not letting too much time get consumed by debates within industry forums. As can be concluded, proprietary standards are sometimes the only option available for companies ready to go to market; yet open standards are the primary way to challenge those early movements. In the end, if one can obtain a de facto standard position in the industry, then it may be less important to hold a proprietary standard. However, customers ultimately not only require early products but demand competitive solutions that are offered from multiple vendors and are interoperable, so a published standard from an industry forum is still heavily sought.

As briefly mentioned, industry forums have a systematic series of events that transpire through the development of a standard that involves numerous iterations between technological capabilities and market requirements.

1. Initially, interest in a particular technology and/or in serving a market is expressed by committee members, and an investigative activity commences.
   - This process can be skipped or could last well over a year
2. Once that interest gains momentum, a study group is formed to determine what potential industry segment and market needs a new standard could serve, as well as whether there is a candidate technology and what its potential features are.
   - With committee members consisting of engineers, architects, product designers, businesspeople, and long-term independent standards consultants, there is typically sufficient motivation to explore the development of a new standard.
   - At the same time, because other standards already exist, there is sufficient motivation by some to delay or eradicate the deployment of new technology standards.
   - As is evident, politics is inevitable in industry forums as commercial success is at stake, both in terms of competing with new technologies and preserving the longevity of existing technologies.
3. A call for applications is then requested whereby usage models and applications are explained, depicting how new market opportunities are created or developed using these new capabilities.
4. After reviewing the applications, market requirements are typically derived and serve as the basis for technical requirements.
5. Historically, once technical requirements have been prepared, a study group has been likely to evolve into an official task group with a specific charter defined and approved.
   - This task group is intended to develop a specification according to a set of metrics that will be published within a certain time frame.

6. In preparation for hearing technical proposals to form the baseline of this endeavor, a series of logistical and procedural documents is created, including technical proposal selection criteria and down-selection procedures.

7. Once proposals are offered to the standards committee, technical evaluation, merging of proposals, elimination of proposals, and peer review help conclude a baseline specification.

8. Once the specification completes the technical due diligence stage and is fully documented, the consensus-driven specification gets committee and oversight approval and is then published.

   - However, politics throughout the process are ever present and interfere with and can derail standards efforts.

   - In an effort to accelerate the development of an open standard, companies offering proposals may elect to work offline to agree on technical differences and possibly compromise to yield a solid proposal.

   - However, companies pursuing a proprietary standard are likely to be unsupportive of modifying their proposals and unwilling to entertain compromise fairly.

10.3 Technological Pioneers in Pre– and Early IEEE Standards

Several companies helped pioneer UWB technology between the late 1980s and mid-1990s and promoted what is today known as impulse radio UWB. They developed the technology for communications, precision ranging, and radar applications, and they advocated the technology to numerous government agencies and commercial corporations. The advantages they claimed were truly astounding, promising very high data rates, very low power consumption, precision ranging capability, high-accuracy radar imaging, multipath immunity, low probability of intercept (LPI), low probability of detection (LPD), spectral reuse, and low-cost implementations. Undisputedly, the technology warranted further investigation because of the many benefits, and since the technology employed new communications concepts, it needed to be demonstrated and also shown to be commercially viable. Potential users of the technology were astounded by the claims, and they immediately set out to define applications that could benefit from these new capabilities, but they were also somewhat skeptical because the technology was fundamentally different from existing wireless technologies. Some thought that UWB technology was not subject to the laws of physics, but, as time would tell, they were.

The reality of UWB technology advanced significantly when, in February 2002, the U.S. Federal Communications Commission (FCC) allocated an unprecedented amount of unlicensed spectrum for UWB communications. Having completed a four-year investigation into UWB technologies, this FCC regulatory milestone helped advance international standards activities and allowed the IEEE high-rate-communications standard to mature, an investigation that had just begun in late 2001. Actually, the IEEE 802.15.3 Study Group a (SGa) committee created a forum in which UWB technology would undergo scrutiny by semiconductor competitors, OEM/ODMs, government agencies, and universities. Prior to the formation of 802.15.3SGa, there were few public disclosures, and technical details were scarce. In early 2002, with an industry specification under development, several companies with their own UWB implementations in hardware, and the promise of commercial success, competitive behavior was ever present. Approximately 40 people made up the IEEE 802.15.3SGa committee in early 2002; they represented major OEMs, semiconductor companies, and small venture-capital-funded companies, and each looked at these standard proceedings in a different light. The OEMs were most concerned with understanding the full spectrum of technological features and how they mapped to new usage models; some of the venture-capital-funded fabless semiconductor companies were primarily focused in getting the standard written so they could enter production; other semiconductor companies were performing due diligence on commercial feasibility with production expected several years later.

UWB developers understood standards to be a critical element in the successful consumer adoption of UWB-based products. As such, proprietary UWB technology innovators, along with major OEMs, ODMs, and semiconductor companies, all engaged in international standardization efforts involving many industry forums. Such industry forums included the IEEE, Ecma International, the Universal Serial Bus Implementers Forum (USB-IF), the Bluetooth Special Interest Group (SIG), the 1394 Trade Association (1394TA), the WiMedia Alliance, the Multiband OFDM Alliance SIG, and the UWB Forum. Each of these forums is dedicated to ensuring that technology is well developed and ultimately certified, is interoperable, and is architected and optimized to exceed the capabilities of existing networking technologies. The first forum in which UWB was discussed was the IEEE because it was there that the physical layer (PHY) and medium access control (MAC) layers were to be specified and from which other standards bodies would build.

Within the IEEE 802.15 Working Group, there are several standards for wireless personal area networks (WPANs), including the technology published as IEEE 802.15.1, which was developed by individuals that make up the Bluetooth SIG. The 802.15.3 standards effort was originally pursued as an evolution to the Bluetooth specification, but the outcome was a standard that was not backwards compatible. This effort was spearheaded primarily as a means to enable fast downloads (up to 55 Mbps) of digital still pictures from cameras to kiosks. The associated PHY and MAC specifications were primarily optimized for file transfers, digital imaging, and multimedia applications. UWB technology emerged as a strong candidate but ultimately was not selected because regulations had not been set. So, in late 2001, the IEEE 802.15 Working Group announced the formation of the IEEE 802.15.3SGa committee. UWB pioneers involved in the early days of this new standards group included General Atomics, Mobilian, Time Domain, Wisair, XtremeSpectrum, and tech-savvy multinational corporations like Eastman Kodak, Intel, Sony, and Texas

Instruments (TI). In the very early days of this committee, Intel presented a concept for Wireless USB (WUSB) [5], and Sony presented requirements for high-speed wireless connectivity of consumer electronic devices. Both General Atomics [6] and XtremeSpectrum [7] directly called out the requirement of 480 Mbps to enable WUSB and Wireless 1394.

Early on, it was clear that UWB technology would enable evolutionary extensions to the ever-popular wired high-speed connections of USB 2.0 (at 480 Mbps) and IEEE 1394 (at 400 Mbps) and possibly take Bluetooth to a whole new level. High data rate, system scalability, small form-factor, low cost, low power, and spectrum flexibility were the key technical characteristics that would evolve short-range wireless communications. Additionally, there were several other high-priority metrics based upon the desire to enable high-speed, fully ad hoc wireless networking. As a result, teams of engineers from competitive companies collaborated to yield a technology that would serve the consumer electronic, personal computing, mobile device, and automotive industries for many years to come.

The IEEE 802.15.3a committee grew steadily once the FCC issued its unlicensed spectrum allocation and continued to grow as technical proposals were disclosed publicly the following year, resulting in over 150 committee members. These two milestones brought this relatively unknown technology into the public eye and allowed for great scrutiny and technical discussion.

With numerous agendas being simultaneously pursued within the IEEE 802.15.3SGa committee, it was to be expected that some companies would want to operate under an aggressive schedule while others would prefer a longer time line. Those pursuing a standard in the shortest amount of time were promoting the proprietary standards option, while the other companies were more supportive of an open standard. This particular committee saw both approaches pursued simultaneously, and they fueled each other. With so many traditional competitors engaged, no one was prepared to yield a competitive advantage, especially time to market and intellectual property claims, so there was much resistance to anyone dominating these proceedings, including the company advocating for a proprietary standard. Since the company pursuing the proprietary standards option was not particularly interested in modifying its solution, it appeared to have isolated itself from the other innovators. Recognizing that a company was pursuing the proprietary standards path, the other UWB innovators started discussions among themselves to identify a means to collaborate for an open standard.

10.4 Standard Efforts of IEEE 802.15.3a

Within the IEEE, XtremeSpectrum, one of the early pioneers of UWB technology for commercial applications, pursued a proprietary standard for UWB by promoting the adoption of its DS-UWB solution, whereas the following four companies were the first major players in the open standards efforts: Discrete Time Communications, which later became known as Staccato Communications with expertise in innovative UWB communications dating back to 1996 through Interval Research and Fantasma Networks; General Atomics' Advanced Wireless Group, which had developed the first multiband technology known as Spectral Keying; Intel Corporation, which had invested in, built, and advocated the technology; and Time Domain Corporation, which had pioneered impulse radio UWB and spun out Alereon.

10.5 Open Standards Efforts

Discrete Time Communications, General Atomics, Intel, and Time Domain Corporation recognized the need to collaborate early on to deliver a winning proposal, so in an effort to help accelerate the down-selection process of IEEE 802.15.3a, they got together in September 2002 to discuss technical commonalities and values. The primary unifying goal was to converge on a single proposal based on UWB multiband technology that would have the largest market applicability and most flexible spectrum capabilities. These UWB innovators had previously developed impulse radios and concluded that in order to meet the industry requirements set forth by consumer electronic, personal computing, and mobile phone companies, a different approach to UWB was necessary; this approach is now known as MB-OFDM UWB. Some of the early companies that supported MB-OFDM actually had given up their own impulse radio proposals and other multiband technical solutions, conceding any competitive lead they may have had in order to contribute to the development of a UWB approach that would become coupled with Orthogonal Frequency Division Multiplexing (OFDM).

By January 2003, additional companies, including Philips and Wisair, banded together with the other four with the common belief that the UWB standard should be based on multiband technology. This informal group worked to refine its united vision, prepared a merged proposal, and repeatedly tested performance. This effort was dubbed by the press the Multiband Coalition (MBC). In an effort to gain additional support, the MBC hosted a cocktail event for the major consumer electronics companies, including Eastman Kodak, Panasonic, Samsung Electronics, and Sony, as well as leading semiconductor corporations, including Infineon Technologies, Motorola, STMicroelectronics, and TI. The MBC presented a convincing argument that changed the fate of UWB. Moving forward, the coalition members recognized that their audience shared a common goal: securing an industry standard that would help produce the best possible PHY specification and afford global regulators comfort and flexibility in their allocations of unlicensed spectrum. Out of that gathering came further cooperation, performance testing, and development efforts from industry participants, with additional members joining the informal coalition.

In March 2003, 26 proposals were presented before the IEEE 802.15.3a standards committee; of these, 25 were based on UWB, and 16 of those proposals were based upon multiband technology. At the meeting, TI presented a multiband proposal based on OFDM that showed available range performance superior to all other proposals at the time. In an effort to create the best possible standard and ensure rapid development, companies from multiband proposals and from several of the single-band UWB proposals entered into a series of technical deliberations outside of the IEEE between March and May. Some members of the MBC collaborated with TI to validate the company's results. As part of that effort, Staccato Communications duplicated TI's system simulation and independently analyzed the system's complexity, presenting the results at the IEEE meeting in May 2003. Further discussions followed, involving all coalition members; finally, at an ad hoc meeting in Denver, Colorado, in June 2003, the decision was made to adopt OFDM as the preferred modulation scheme for multiband UWB. In order to better reflect the evolution of the multiband movement, the MBC was reborn as the Multiband OFDM Alliance (MBOA) in June 2003. At that time, Hewlett-Packard, Microsoft, NEC Electronics, Panasonic, Samsung, STMicroelectronics, SVC Wireless, TDK, TI, and Wisme joined existing coalition members Femto Devices, Focus Enhancements, Fujitsu, General Atomics, Infineon, Institute for Infocomm Research, Intel, Mitsubishi Electric, Royal Philips Electronics, Staccato Communications, Taiyo Yuden, Time Domain, and Wisair.

At the July 2003 IEEE meeting, member companies of the MBOA and Sony Electronics submitted a proposal for a multiband OFDM–based approach as the IEEE's 802.15.3a PHY specification, known formally as Merged Proposal #1. While their backgrounds were varied, including consumer electronics, digital imaging, home entertainment, personal computing, and semiconductors, MBOA members shared commonality in designing and building systems and silicon based on UWB. Meanwhile, though invited to participate in these discussions, XtremeSpectrum excluded itself from such mergers and continued to advance its proposal, declaring technology superiority and time-to-market advantages as it continued its proprietary standards efforts.

10.6 Proprietary Standard Efforts

XtremeSpectrum helped form and chaired IEEE 802.15.3SGa and authored a mature technology proposal for impulse radio, referred to as DS-UWB. XtremeSpectrum, a fabless integrated circuit startup company that had been developing UWB technology since 1999, was later acquired by Motorola SPS and is now part of Freescale Semiconductor. XtremeSpectrum first announced commercial products in 2001 and clearly held a competitive advantage over other UWB developers in late 2001 and 2002. XtremeSpectrum was positioned very well within the IEEE standards committee and also started engaging with the 1394TA in an effort to introduce Wireless 1394. The company declined invitations to preview other UWB proposals because its mature proposal, if the 802.15.3a schedule went unchallenged, would have been the best baseline candidate.

To counter the momentum of the multiband-based proposal, XtremeSpectrum presented a number of arguments against the multiband proposal, including technology immaturity, interference concerns, noncompliance with FCC regulations, complexity issues, and power-consumption worries. The company could rightly criticize the multiband proposal for not being fully formed in comparison with its own proposal, which had been committed to silicon and was supported with simulations. Motorola SPS conducted numerous experiments and reported that there were sufficient interference issues to preclude multiband technology from gaining industry acceptance and, more importantly, regulatory certification; on the other hand, the Motorola proposal was undisputedly compliant with the FCC Report & Order. Lastly, the only disputes that remain today have to do with complexity and power consumption; complexity will be determined by solution pricing, and power-consumption claims can be substantiated when each technology gets into production.

Despite the large number of companies unsupportive of the Freescale proposal within the IEEE, Freescale was unwilling to surrender its time to market and other advantages and continued to promote its technology in the OEM/ODM space. The company tapped new markets, including China, and believed that if it could gain acceptance by just a few dominant companies and get its solution out rapidly, then it could earn the de facto standard position and positively influence the IEEE or, for that matter, completely make those proceedings irrelevant.

10.7 Concurrent but External Standards Efforts to the IEEE

At the September 2003 IEEE meeting in Singapore, IEEE 802.15.3a Merged Proposal #1 came very close to achieving the 75 percent approval required to conclude down-selection, but technical due diligence of the MB-OFDM proposal, as well as politics within the committee, prevented the baseline from being established. Merge Proposal #2, which is based on the XtremeSpectrum original proposal, remained the minority for months and months. Over the next half-year, the technology of Merge Proposal #1 gained incredible industry momentum through technology maturity and a series of endorsements and engagements, but it never obtained the necessary 75 percent IEEE approval to advance beyond the down-selection stage.

Concurrently with the IEEE standards efforts, in an effort to advance the technology without abandoning the IEEE, both proposals pursued external acceptance by means of industry alliances and forums. They both grew their memberships and encouraged industry adoption, but they maintained their difference of one being an open standard and the other a proprietary standard. The open standards efforts saw the Multiband OFDM Alliance become a legal entity and then merge with another open forum (the WiMedia Alliance) and ultimately gain endorsements and recognition from the USB-IF, 1394TA, and Bluetooth SIG. Meanwhile, the DS-UWB efforts advanced within the UWB Forum and gained recognition from global participation. The fundamental differences between the two organizations are the caliber of membership and the openness of the specifications. WiMedia boasts paid membership from most top-tier personal computing, consumer electronic, mobile phone, software, and semiconductor companies, as well as hundreds of millions of dollars in startup venture-capital financing. Further, WiMedia's specifications are drafted through consensus-driven technical discussion with a technical oversight committee and a board of directors that includes ten world-class companies and three leading start-ups.

Little is known about the UWB Forum except that its board of directors includes Motorola, Freescale, and Pulselink, a California-based startup company, and that membership at lower tiers appear to be free (at the time of publication, Freescale and Motorola have withdrawn from the UWB Forum). The UWB Forum has setup branches in the United States, Japan, and China and has spoken about performing product certification. No schedules have been released regarding specifications, and the only specification suspected to have been written is the PHY. The UWB Forum does discuss a common platform similar to that introduced by the WiMedia Alliance as the UWB Forum would like to serve proprietary WUSB, the 1394 Trade Association, as well as Bluetooth and IP. Meanwhile, the WiMedia Alliance has introduced three specifications that make up its common radio platform (PHY 1.1, MAC 1.0, and MAC-PHY Interface 1.0) and has announced compliance workshops for 2006 and a schedule to get certified products out in time for holiday season 2006. Further, WiMedia is now working on next-generation efforts for both the PHY and MAC to meet the needs of the evolving marketplace. At least six companies to date have announced products based on the WiMedia specifications, while only Freescale has developed silicon based upon DS-UWB.

The WiMedia Alliance was formed to serve the growing high-rate WPAN industry in a fashion similar to how the WiFi Alliance serves IEEE 802.11. WiMedia is the forum for ensuring compliance and interoperability and provides a common radio platform for upper-layer protocols, including Certified WUSB, Wireless 1394, Bluetooth over UWB, and IP over UWB. WiMedia originally supported the 2.4 GHz PHY and centralized MAC of IEEE 802.15.3, but with little interest in the IEEE 802.15.3 specifications and persistent deadlock in 802.15.3a, the WiMedia Alliance departed from the IEEE and adopted the MBOA MAC and PHY specifications in May 2004 [8].

The 1394TA had its eyes on high-speed wireless for some time and had developed, but did not publish, a protocol adaptation layer (PAL) for 802.11. In 2004, the 1394TA published a PAL for the IEEE 802.15.3 standard. This standard, though using a 2.4 GHz PHY, was based on the 802.15.3 MAC, so UWB developers, including XtremeSpectrum, saw a PHY migration path that would allow them to take advantage of the completed PAL and deliver even higher performance with a UWB PHY. In September 2004, the 1394TA announced that it was now in collaboration with both the WiMedia Alliance and MBOA for Wireless 1394 [9]. In practicality, likely only one Wireless 1394 solution will ever really materialize; otherwise, interoperability and branding could yield consumer confusion.

The WUSB Promoter Group, backed by Intel, Microsoft, Philips, Agere, NEC Electronics, and Hewlett-Packard, including key contributors Alereon, Appairent Technologies, Staccato Communications, and Wisair, adopted the MBOA specifications in February 2004, and also engaged with the WiMedia Alliance. The WUSB Promoter Group introduced the Certified WUSB specification and handed it over to the USB-IF, which manages the widely successful wired USB industry and issues compliance logos. The Certified WUSB specification of the USB-IF is a protocol that was architected and optimized for wireless connectivity, including security and association, power management, data throughput, bandwidth allocation, and operating system driver reuse.

Building on the expectations of UWB, the technology must be designed to support much more than just wireless extensions to USB and 1394 (a.k.a. FireWire). In January 2004, the MBOA took on the task of investigating a new MAC, known today as the WiMedia MBOA MAC. Actually, this effort originated out of a group of seven multinational corporations (CE7), including leading consumer electronic companies, that believed the 802.15.3 MAC was insufficient to provide the capabilities they envisioned for UWB-based applications. The CE7 included Hewlett Packard, Panasonic, Philips, Samsung, Sharp, Sony, and Toshiba. They went so far as to inform silicon vendors that despite their interest in UWB technology, without a decentralized MAC that could offer mobile ad hoc networking with good quality of service (QoS) and support for many simultaneous collocated systems, they simply would not design in their silicon. The MBOA MAC was an intensive project that managed to achieve its aggressive schedule due to the strongly supportive companies that assigned diligent architects and engineers devoted to the cause and felt passionate about developing a revolutionary wireless system. Meanwhile, within the IEEE standards forum, a revision to the 802.15.3 MAC was also under way to improve the standard and add some functionality, primarily advocated by members of the UWB Forum and Merge Proposal #2.

The IEEE 802.15.3a committee continued to meet every two months, and a face-off remained between the MB-OFDM-based proposal #1 and the DS-UWB-based proposal #2. Both sides attempted to compromise, but nothing ever materialized. Namely, the DS-UWB advocates claimed (and lobbied the FCC) that MB-OFDM technology was in violation of the FCC rules for UWB; similarly, the MBOA proclaimed its superiority in spectral shaping to allow for a cognitivelike radio. Furthermore, the DS-UWB fabless semiconductor company XtremeSpectrum, whose assets now belonged to Motorola SPS, continued to demonstrate its technology and product maturity, while finding technical faults with the MB-OFDM approach. As it turned out, some of the claims were proven incorrect, and the substantial technical due diligence resulted in revisions to the MB-OFDM proposal that made it stronger. Here, the true essence of technical peer review had directly and positively affected the outcome of the MB-OFDM technology. Unfortunately, many details were not fully explained regarding Merge Proposal #2; therefore, a less thorough technical peer review transpired for DS-UWB.

With no end in sight to the deadlock within the IEEE 802.15.3a standards committee, and with increasing politics and new voters, the MBOA announced that it had formed itself as a SIG that would allow for the publication of its own specifications, inclusive of the PHY, MAC, and MAC-PHY interface specifications. Now, with the MBOA positioned to deliver the MAC and PHY to the WiMedia Alliance's common radio platform in a more deterministic time frame, and having been selected by the WUSB Promoter Group and involved with the 1394TA Wireless Working Group, the standards efforts were well en route to being finalized, though the forum had changed from that of the ISO accredited IEEE. The notion of needing SIG activity in parallel with the IEEE is not new; in fact, the Bluetooth SIG was formed in order to develop and accelerate specifications for the first short-range wireless standard. In fact, the Bluetooth SIG gave the technology back to the IEEE, and it is now published as IEEE 802.15.1. Recognizing the need to show strong support and a means to certify products, in late 2004, the UWB Forum was formed to support DS-UWB PHY technology and the IEEE 802.15.3 MAC [10].

In March 2005, in recognition of the complementary nature of the MBOA-SIG and the WiMedia Alliance, and in order to increase efficiency and reduce confusion, they merged and are today known as the WiMedia Alliance, whose offerings include a common radio platform (comprising the MBOA MAC, MAC-PHY interface, and PHY and also supplemented with policies that allow for coexistence and collocation of multiple protocols residing atop the common radio platform), a protocol stack for IP, a global regulatory committee, and a compliance and interoperability effort.

On May 4, 2005, in an effort to perpetuate the Bluetooth technology into the future, the Bluetooth SIG announced a roadmap for High-Rate Bluetooth based on UWB technology and formed a study group initiative to explore high-rate requirements and to evaluate both WiMedia's technology and DS-UWB supported by Freescale Semiconductor (this was Motorola SPS until it spun out) and others involved in the UWB Forum [11].

In a further effort to appease the many consumer electronic companies involved with the WiMedia Alliance, and with the longevity of this standard expected, WiMedia announced in August 2005 that it was actively working with the ISO-accredited international standards body known as Ecma (Ecma is most widely recognized for standardizing DVD technology) to deliver a published standard for the WiMedia specifications by the end of 2005 [12]. While this path was pursued, the IEEE standards opportunity remained a possibility, as did the 5 GHz OFDM technology for both IEEE 802.11a and ETSI's HyperLAN. In December 2005, Ecma-368 and Ecma-369 were published by Ecma International and are downloadable from http://www.ecma-international.org/publications/standards/Standard.htm. In the second half of 2005, development kits, reference designs, and preproduction products were introduced, and compliance and interoperability became the new craze as issuance of the Certified WUSB and WiMedia logos grew closer. Alereon, Staccato Communications, and Wisair demonstrated interoperable radios in December 2005, building confidence that UWB-enabled consumer products are on track for 2006.

The Consumer Electronic Show (CES) held in Las Vegas, Nevada, January 2006, was a major event for UWB technology developers. Nearly a dozen WiMedia member companies introduced products and/or performed demonstrations of Certified Wireless USB, Wireless Streaming Video, or Bluetooth over UWB. CES 2006 was also where Freescale chose to announce several design wins; they introduced their pre-paired, closed-system, cable-free USB solution and demonstrated automotive wireless multimedia. The show featured numerous form-factor demos that are expected to become consumer products in late 2006. Shows throughout the year will continue to build on this theme, which will ensure a steady introduction of consumer products in late 2006, ramping up through 2007.

It is worth noting is that XtremeSpectrum's constant opposition to the MB-OFDM effort caused great unification among traditional competitors within the WiMedia Alliance, helped further the MB-OFDM specifications, and actually helped ensure that corporate resources and funding were allocated to conduct interference tests and perform extensive simulations on the technology.

10.8 Another Open Standard: IEEE 802.15.4a (Low Bit Rate)

Another IEEE standards effort was started for UWB technology, but this one was focused on low-bit-rate applications. The effort, known as IEEE 802.15.4a, started in November 2002 to define an evolutionary set of capabilities for the merging market of Zigbee and Active Radio Frequency Identification (RFID). Here, the market is industrial controls, home and factory automation, and asset tagging and tracking. The primary metrics are low cost, long battery life, precision ranging, and multipath immunity. Impulse radio UWB is the primary technology candidate, and the specification is slated to be published in early 2007. The standards committee, having experienced the deadlock and politics of IEEE 802.15.3a and WiFi standards IEEE 802.11g and 802.11n, has elected to take a different approach to standardizing the technology; it will build in optional modes to avoid contention and deadlock—basically, multiple PHYs. It is not yet known if this approach will yield standards-based products more quickly than the other approaches or if it will merely postpone the inevitable decision to select a single PHY and, as a result, actually cause delays in market adoption while it identifies the single solution. Further, it is not known if the overall standard will suffer in its creation by having too many optional components, limiting the amount of technical due diligence and review of each section by the committee. At present, substantial objections within the committee persist because of the many incompatible optional modes and the need to coexist with high-bit-rate UWB solutions. For more information on the IEEE 802.15.4a standards activities, one should visit www.ieee802.org/15/pub/TG4a.html.

10.9 Conclusion

The WiMedia technology is on track to be introduced to consumers in late 2006 as Certified Wireless USB–enabled products, and Freescale is expected to have some uptake on DS-UWB for wireless video streaming and in some vertical markets. In 2007, WiMedia is expected to have IP connectivity to allow for high-speed peer-to-peer communications, and this may be promoted as WiMedia's WiNet, Digital Living Network Alliance (DLNA), Universal Plug and Play (UPnP), and/or Bluetooth PAN. It is the author's belief that WiMedia's open standards–based technology will become the preferred high-speed wireless solution due to the competitive landscape of solution providers, the immense industry support of consumer electronics, personal computing, mobile phone, and automotive companies, the overall capabilities of the technology, and the recently defined efforts to deliver revisions and next-generation specifications. Further, the author believes that if DS-UWB technology had been able to get to market in 2003, then it would have had a good chance of becoming de facto, but delays in production and the evolution of market requirements are unforgiving.

References

1. See http://grouper.ieee.org/groups/802/11/Documents/DocumentArchives/1990_docs/1190010.doc.
2. See www.dolby.com/about/who_we_are/history_1.html.
3. See http://standards.ieee.org/sa-mem/why_std.html.
4. See www.iso.org/iso/en/aboutiso/introduction/index.html.
5. 02139r0P802-15_SG3a-Intel-CFA-Response-Wireless-Peripherals presented by Chuck Brabenac to the IEEE 802.15SGa committee in St. Louis, Missouri, March 2002.
6. 02143r0P802-15-SG3a-Application-Opportunities-GA presented by Dr. Roberto Aiello, Jason Ellis, and Larry Taylor to the IEEE 802.15SGa committee in St. Louis, Missouri, March 2002.
7. 02031r0P802-15_SGAP3-CFAReaponseAltPHY presented by Pierre Gandolfo to the IEEE 802.15SGa committee in Dallas, Texas, January 2002.
8. See www.multibandofdm.org/press_releases.html.
9. See www.1394ta.org/Press/2004Press/september/9.7.a.htm.
10. See www.uwbforum.org/index.php?option=com_content&task=view&id=28&Itemid=69.
11. See https://www.bluetooth.org/admin/bluetooth2/news/story.php?storyid=512.
12. See www.wimedia.org/en/events/index.asp?id=events.

11

Commercial Applications
by Roberto Aiello

UWB is uniquely suited to enhance some of the most popular commercial products, such as PCs, printers, digital cameras, media players, and external mass-storage devices, with wireless connectivity or wireless personal area network (WPAN) functionality. A WPAN is defined as a network of personal devices, usually connected at short range. It is different from a wireless local area network (WLAN), which is an extension of the network infrastructure, or a wireless wide area network (WWAN), which is intended for metropolitan coverage, including roaming and mobility. Industry experts project that the first markets for UWB will be driven by the PC and its peripherals and will begin with aftermarket add-ons, with native UWB implementations expected to follow. Many expect that this will lead to quick adoption of UWB by mobile device manufacturers who will want their devices to connect to the enabled PCs. The market is also expected to find traction in cable-replacement applications for home multimedia audio-video applications, including video cameras.

Silicon is already in production to enable Certified Wireless USB (WUSB), which will be part of the first wave of UWB implementations for commercial applications. The WUSB Promoters Group (www.usb.org/developers/wusb) has certified numerous WUSB profiles to ensure interoperability of WUSB products. Some of the first integrated circuits to enable commercial WUSB connections are expected to be introduced shortly. As ad hoc and peer to peer applications become more popular, Internet Protocol (IP) is expected to grow in popularity as an alternative to WUSB. Chip sets have also been developed by Freescale Semiconductor based on a different technology for replacing cables for computer peripheral, video, and home multimedia markets.

This chapter covers the potential commercial opportunities for UWB, including the UWB value proposition in these markets, examining some specific target markets and market challenges.

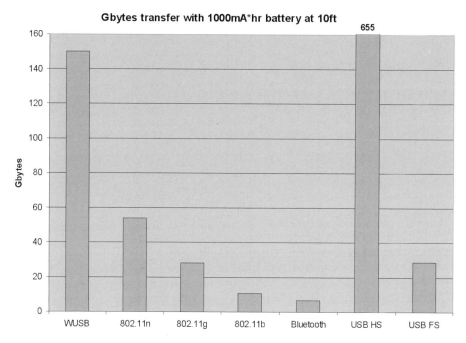

Figure 11.1 Power-consumption comparison. USB power consumption is assumed to be 800 mW in active mode (transmit and receive) running at 480 Mbps.

11.1 UWB Value Proposition

The main commercial applications for UWB are for high bit rate communication at short range. This is because of its spectrum characteristics, as shown in Chapter 1, because, on the one hand, the large bandwidth available makes is easier to achieve high bit rate communications; on the other hand, the low transmit power makes is difficult to achieve long ranges. This has led the industry to focus on WPAN applications.

As a result, the fundamental value proposition of UWB, compared to other technologies that use other parts of the spectrum, is based on its high bandwidth combined with high efficiency per bit in terms of power consumption. As a wireless technology, UWB is expected first to find its market as a cable replacement for wired technologies like high-speed USB 2.0. In this case, the most compelling use would be for connecting mobile devices to other mobile devices, such as a digital camera to a portable media player (PMP), because there are no existing alternatives. Other connections for fixed devices, such as PC to printer or digital video camcorder to PC, are also very interesting.

A comparison of energy efficiency is shown in Figure 11.1, where both wired and wireless technologies are considered. The plot shows how many gigabytes can be transferred with a 1,000 mA × hr battery at a 10 ft. distance. UWB-based systems are shown to be more efficient than other wireless technologies. The other technologies considered are optimized for other application requirements; 802.11x, for example, operates at longer distances and does not compare as well at a 10 ft. distance. It must be noted that UWB would perform poorly if the comparison were done over a longer distance.

To succeed, the UWB market will need to find creative new applications or superior functionality as compared to alternative access technologies. Once compatible forms of UWB are available in a variety of high-volume devices, this will drive additional uses and consumer uptake since basic connectivity will already exist.

11.2 Potential Markets for UWB

Early adoptions of UWB in volume will include computing and consumer devices for use in residential and mobile applications. Devices in these segments will use either WUSB, IP, or Bluetooth as the access protocol.

There are no high-speed WPAN products on the market yet, so there is limited feedback from consumers about the types of applications likely to be most successful. However, most PC, consumer electronics, and mobile companies are of the opinion that there is a potentially large opportunity to distribute media between fixed and portable devices, such as digital cameras, digital video camcorders (DVCs), camera phones, MP3 players, portable media players, external hard disk drives (HDDs), personal digital assistants, and desktop and notebook PCs, as shown in Figures 11.2, 11.3, and 11.4.

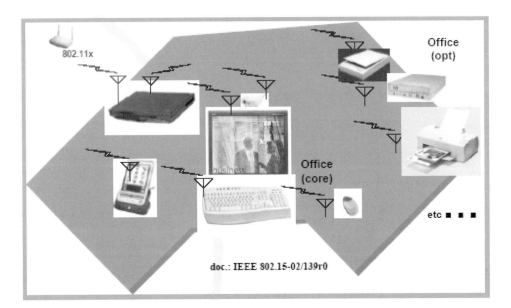

Figure 11.2 Intel's view of applications for UWB. The PC is the center of the connectivity on the desktop. Note the connection of the desktop with the 802.11 network.

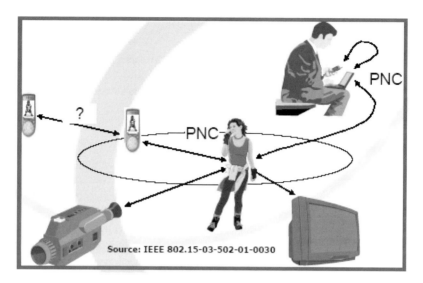

Figure 11.3 Nokia's view of applications for UWB. The handset is at the common platform that requires connectivity to other personal devices.

Applications scenarios include several environments and usage models. In the enterprise, computers can exchange data when in the same conference room; they can use local disk space as virtual memory, for synchronization, or for data backups; and laptops can connect to LCD projectors. In the home, digital devices can stream audio and video to speakers and TVs; gaming devices can directly connect to each other without need for an infrastructure; printers, cameras, and other peripherals can connect ot a PC or a set-top-box. In public spaces, people in transit can exchange video clips or MP3 audio files, or they can buy CD or DVD contents from kiosks.

In general, consumer demands for wireless data can be broken down into three key areas:

- Moving data. This includes data from source to memory, from memory to destination, or from memory to memory. These tasks involve stationary, mobile, and peer-to-peer operation. This is the main connectivity application envisioned for WPAN. Examples of source devices are DVD, PC, or MP3 players (when connected to speakers). Examples of memory include hard drives or solid state memories, and examples of destination devices include displays or MP3 players (when connected to a PC).
- Sharing data. This includes connecting with local infrastructure or fixed networks and client/server environments, as well as connecting via shared access points. These tasks generally involve WLANs.
- Seamless operation. This includes roaming coverage, accessibility, and being always connected. This is mostly an operation for WWANs.

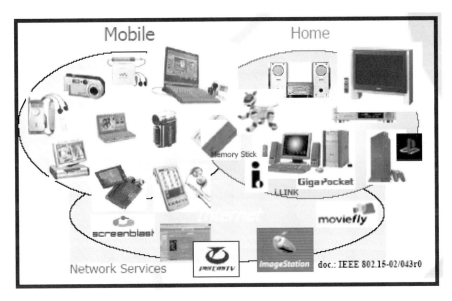

Figure 11.4 Sony's view of applications for UWB. Mobile, computing, and entertainment devices require peer-to-peer connectivity with each other.

The trend continues for users to want to transfer larger files. This is driven by both the availability of larger storage and the demand for richer content. For example, an MP3 version of Bob Dylan's "Blowin' in the Wind" is approximately 4 MB for 3 minutes of audio, while an H.264 or MPEG4 version of the TV show *Lost*, available today for the iPod, is about 200 MB for 40 minutes of video. At the same time, high-end mobile terminals currently have up to 5 GB of on-board hard disk memory. The ability to transfer large media files between multiple portable devices or between a portable device and a server, kiosk, or display is an attractive option for consumers. This means that data throughput requirements for WPANs have grown substantially, and they will continue to grow in the future. UWB is expected to meet these requirements and will need to continue to adapt as demands evolve. The rise of camera phones and the popularity of digital cameras are also leading to an increase in video clip exchanges from device to device or from device to media (such as a CD or DVD).

While UWB is capable of addressing a wide variety of these and other applications, the initial and early volume devices will likely center on PCs, digital cameras and media players, printers, and external mass storage.

Future applications may include human interface devices (mouse, keyboard, gamepad), real-time displays (TV, monitor), and other consumer digital multimedia devices, such as set-top boxes and home-theater equipment. Figure 11.5 depicts the major classes of UWB-enabled devices. The following sections consider in detail each of these four classes.

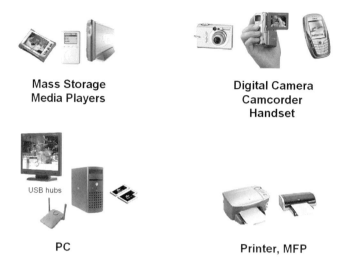

Figure 11.5 The devices that will benefit from UWB include PCs, digital cameras, printers, and mass storage. Each class of device has specific characteristics and requirements.

11.3 PC Applications

Because the PC takes many forms, it provides a wide variety of usage models and applications. Figure 11.6 shows some of the peripherals to which a PC can connect. Generally, PCs are either a desktop implementation that stays in a fixed location or a notebook that is mobile. (A docked notebook can be treated as a fixed device similar to a desktop PC.) Traditionally, PC peripherals have been connected using USB, Bluetooth, or WiFi.

Some of the most compelling PC applications for UWB include viewing pictures and video from a notebook PC on a TV, performing dial-up networking with a mobile phone, transferring files between PCs, downloading pictures from a camera to a PC, transferring pictures and audio to a PMP from a PC, and archiving onto mass storage.

11.3.1 Enabling PCs with UWB

In the computer market, UWB capability is expected to be sold initially as an add-on aftermarket solution, either as an external USB dongle for desktop PCs or ExpressCard™ for notebook PCs. The advantages of this solution are that it provides easy installation to consumers and the antenna is in a favorable location. USB dongles, however, are limited in total throughput because of the inherent limitations of the USB bus. Peripheral Component Interconnect (PCI) and Peripheral Component Interconnect Express (PCIe) implementations may be desirable for some applications where maximum data throughput is needed, and an external antenna can be

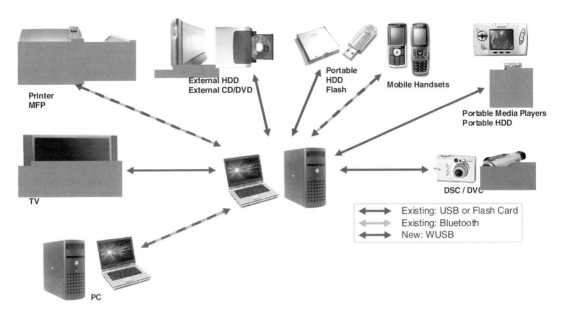

Figure 11.6 UWB offers some compelling applications for linking PCs to peripherals.

implemented properly. This is not that relevant when only few devices are connected to the PC because the bit rate is sufficient to enable the application, but it is relevant when multiple devices are connected at the same time and require a larger aggregate throughput.

11.3.2 PC Applications

Figure 11.7 depicts an external USB dongle for a desktop PC, which supports both USB high speed and full speed. Products like these offer vendors the quickest time to market for UWB applications on PCs.

Alternatively, the same function can be implemented within a USB hub monitor display or monitor stand. In this type of application, the monitor or stand essentially becomes a "personal kiosk" that is conveniently located above the table to maximize the RF reception with antennas located in the monitor itself. Given that such a hub monitor or stand may have several USB ports, a UWB connection could be integrated with a monitor or even bundled with the monitor for sale (Figure 11.8).

Figure 11.7 An external USB dongle to enable UWB connectivity for a PC.

Figure 11.8 An entire UWB dongle can be implemented in a USB hub monitor display or a monitor stand to enable UWB connectivity for a PC.

For new notebook PCs with UWB capabilities, the earliest products will probably be PC Card (using the parallel Cardbus 32 interface) and ExpressCard implementations (Figure 11.9), because they are easy to add by the end user. In this case, the ExpressCard implementation is essentially a USB dongle in an ExpressCard form factor.

11.4 Camera and Media Player Applications

Figure 11.10 shows examples of connection options for digital still cameras (DSCs), digital video cameras (DVCs), and camera phones, including connections to PMPs. Some blending of functionality has begun in the camera market; for example, some DSCs and many new camera phones have MPEG4 video capability, while DVCs can take high-resolution still images.

In a cable replacement application for cameras, the UWB functionality is expected to enable all existing functions between two wireless nodes, each equipped with a radio and associated software. Some first-level applications here include using UWB to print pictures, downloading them to a PC, or burning them onto a DVD or CD. Beyond cable replacement, some of the more interesting applications for

Figure 11.9 PC Card and ExpressCard implementations of UWB functionality enable notebooks.

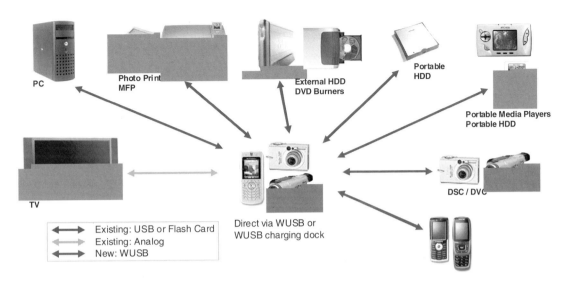

Figure 11.10 Common connections and new potential connections enabled by UWB for digital cameras.

UWB in a camera include sharing pictures between cameras, displaying pictures or video on a TV, PC, or the LCD screen of a mobile device, or downloading the pictures to a portable media player's HDD or memory card.

The memory capacity of DSCs has grown tremendously since their initial introduction, and digital camera market research conducted by top DSC suppliers (e.g., Canon, HP, Kodak) indicates that many consumers have taken to storing their favorite pictures on their camera's high-capacity memory card, using it like a portable photo album to share pictures and video with friends and family. This has led to an intriguing new application for UWB, linking DVCs, DSCs, and camera phones to support the transfer of pictures between these portable photo or video albums.

This particular application is a high-value proposition for UWB because there is no other technology positioned to support this feature. Infrared and Bluetooth are not fast enough to do quick bursting of pictures and videos, and WiFi is currently considered too difficult by most consumers for ad hoc networking. Since this application requires each camera to have this capability, it ultimately provides value to OEMs to sell more cameras. All individuals who want to share pictures or video must have this capability on both ends of the link, and the value builds increasingly with the number of nodes deployed.

Another potential application for UWB is the transference of images from a digital camera to a PMP, a device that is quickly growing in popularity among consumers. Early products introduced in this market include Archos products and one based on Microsoft's Portable Media Center platform. PMPs can integrate an HDD (4 to 80 GB typically) and/or flash memory (128 MB to 1 GB typically).

Transferring pictures and videos from a camera to a PMP also allows for the sharing of personal content from one personal photo/video album to another, where it can become part of another user's personal collection. The usage model is similar to that of transferring files to another camera and to an HDD. UWB is well positioned to enable this application; alternatives for consumers include using a USB cable or a Flash card slot (if the memory cards between devices are compatible).

UWB can also be used to download images to external HDDs (fixed and portable). A portable HDD is typically a 4–80 GB capacity drive that either can be powered off the USB cable when connected to a PC or has its own rechargeable battery. The value proposition for UWB is good for both fixed HDD applications and for portable HDD. (For a fixed HDD, a short cable can be attached for connectivity to a camera. However, for portable HDD it is very inconvenient to be carrying around unconnected cables.)

Perhaps one of the more interesting opportunities for using UWB links with digital cameras is to view pictures or video on a TV or PC. Most DSCs and DVCs are capable of displaying pictures and video on TVs by connecting directly to the composite video RCA jack (or S-video and stereo audio for DVCs) on the front or rear of a TV. This feature is inconvenient and cumbersome for both DSCs and DVCs and in doesn't exist in camera phones. As an alternative, some higher-end plasma, LCD, or microdisplay TVs now incorporate Flash card readers and a JPEG (and sometimes also an MPEG4) decoder so that the Flash card can be inserted to display photos or video, but the process is not user friendly.

This application can be addressed by adding UWB functionality to the TV and camera. The TV will also need to have internal memory (or a Flash card) to store a transferred picture or video file. By enabling this wirelessly, the camera can be located near the viewers and away from the TV screen, allowing it to be used to control the slide show or video. This also allows multiple cameras from viewers to share the screen easily from the viewing area without plugging and unplugging multiple analog cables.

The value proposition for UWB in this application is high. End users have a strong desire to share their photo albums in a home environment, and the current method of using multiple analog cables is cumbersome and unsightly. And, UWB functionality would enable wirelessly controlled viewing, which could not be provided by simply inserting a Flash card into the TV.

11.4.1 Enabling Cameras with UWB

Given the high-value proposition of connecting a camera to another mobile device, a peer to peer functionality must be included. This can be realized either with WUSB, including both "host" and "device" capability in each camera, or with IP, which allows ad hoc connectivity more easily. However, in some applications, it is possible that only the device support will be required (i.e., for printing or file transfer to a PC).

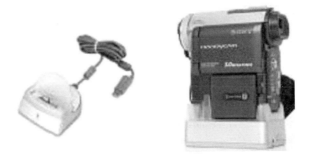

Figure 11.11 Enabling a camera's charging dock with UWB quickly adds connectivity to a printer, PC, or TV.

11.4.2 Camera Applications

External adapters do not work well for cameras, which really require fully integrated UWB capability. One alternative is to enable a charging dock with UWB capability since cameras can be set down for printing. In this type of application (Figure 11.11), the charging dock provides power from the wall. This type of dock could enable UWB connections to a printer, PC, or TV.

11.5 Printer and All-in-One Printer Applications

With the printer or multifunction printer (MFP) in the center, Figure 11.12 shows various products that it can connect to using UWB. Potential applications include all activities that consumers will engage in with printers and MFPs in the home or while mobile.

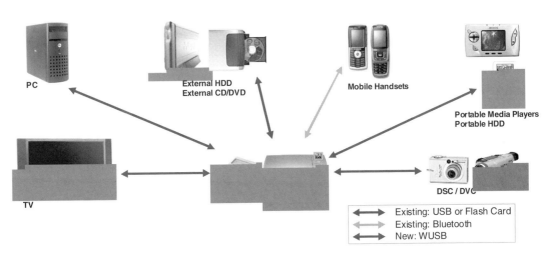

Figure 11.12 There is a range of existing and potential connections for a UWB-enabled printer.

Beyond the convenient cable-replacement applications for UWB in printing, one of the more interesting printing applications is the ability to print from a TV to printer. The only alternative to UWB for this application would be to use Flash card slots in a TV and printer, which immediately raises compatibility, as well as convenience, issues.

This application is especially interesting when coupled with using UWB to view pictures on a TV. Printing from a TV allows an individual end user or a group of people to view pictures together on a large display and then selectively print those of interest.

A similar application is to print directly from a DSC, DVC, or camera phone to a printer. The alternative to UWB in this case is to connect via USB cable or to use Flash. In terms of standard cable-replacement applications, UWB can also make it possible to print directly from a mass-storage device or PC directly to a printer.

11.5.1 Enabling Printers with UWB

Most MFPs and printers are plugged into the wall and can supply sufficient power for high-speed UWB. A growing class of portable photo printers is battery powered, and since the electromechanical print heads only require 4 to 8 Mbps max for high-quality color photo printing, the range and duty cycle can be limited using a higher UWB throughput mode to reduce current consumption significantly.

11.6 External Mass Storage

Long a staple for the enterprise market, external mass storage is now growing in popularity in the consumer market. Users are archiving files into a fixed or portable repository from various devices that capture, download, and/or create the content. Figure 11.13 shows links that could be supported by UWB for mass storage. Beyond standard cable replacements, the new connections that UWB could enable include links between TVs and PMPs and mass-storage devices.

For example, using UWB, consumers can link their TVs to mass-storage devices to view pictures, home video, or stored digital movies. In terms of cable replacement applications, UWB could be used to transfer or archive files from a PC to a mass-storage device, such as a HDD, CD, or DVD, or it could be used to print directly from a mass-storage device. UWB would also be useful to transfer files between a PMP or camera and a HDD, CD, or DVD.

11.6.1 Enabling Mass Storage with UWB

For mass storage, early UWB HDD applications will likely use a daughtercard integrated into the HDD design. Alternatively, an external card could be used, but it would require an external power supply. For portable HDD, only internal implementations will be acceptable.

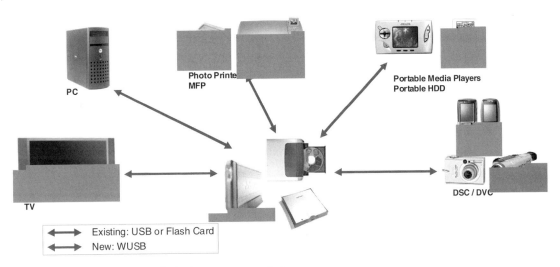

Figure 11.13 Potential UWB links for mass-storage devices.

11.7 Future Applications and Trends

This chapter has shown that UWB has the opportunity to provide an "infrastructure" for both current and future applications. On the one hand, it allows for more convenience for consumers to replace existing cables in existing applications; on the other hand, it allows for creation of new applications not possible today because of the constraints of existing topologies.

After it becomes established in its earliest market applications, some interesting future possibilities and challenges for UWB may lie in handling streaming digital video, which currently requires multigigabit throughput for uncompressed signals on an HDTV.

We are in the midst of an ongoing trend for increased disk capacity and speed in consumer devices. Hard disk capacity on mobile platforms quadruples approximately every 18 months, and networking speed typically grows fivefold every three to four years. For example, from 1995 to 1999, data rates jumped from 1 Mbps in 802.11 to 11 Mbps in 802.11b. Then, from 1999 to 2002, networking speed advanced to 54 Mbps with 802.11 a/g. From 2002 to 2007, it is projected to grow to between 500 and 700 Mbps with 802.11n. These trends in increasing capability underscore the need for UWB products to stay ahead of the projected curve in order to be successful and gain enough market traction and uptake.

Undoubtedly, the pressure to increase data rates will persist over time, and there is a need for an integrated industry strategy to support higher speeds. So, the question is, how do the current roadmaps for UWB measure up?

We have seen companies announcing 1 Gbps products based on proprietary technologies in 2005 and 2006. Applications will require even higher throughput to keep up with the evolution of storage technology. We should expect companies and standards organizations to consider multiple-gigabit-per-second throughputs in the next few years. The good news is that most of the current UWB designs and implementations utilize only 20% of the spectrum available in the United States (1.5GHz out of 7.5GHz available). This leaves a lot of room for improvements and evolution of the technology. Obviously, it is technically challenging to move to higher operating frequencies and higher bit rates; however, the industry should be able to use current discoveries to push technology forward.

As UWB finds its stronghold in the wireless marketplace, its supporters and developers will need to continue to focus on simplifying use, streamlining costs, and staying ahead of advancing data rates.

11.8 Acknowledgments

The author would like to thank Jarvis Tou for many inspiring and educational discussions and for having produced many of the pictures in this chapter.

About the Contributors

Roberto Aiello is the founding CEO and now CTO of Staccato Communications. Prior to working at Staccato, he was founder, president, and CEO of Fantasma Networks, an ultra wideband (UWB) product company. Previously, Dr. Aiello led the wireless research and built the first documented UWB network at Interval Research, Paul Allen's research laboratory. Earlier, he held senior positions at the Stanford Linear Accelerator Center and the National Superconducting Super Collider Laboratory in Texas. Dr. Aiello is a recognized leader in the UWB community, and his efforts were instrumental in getting UWB spectrum allocated in the United States. Dr. Aiello holds a PhD in physics from the University of Trieste. He serves on several advisory boards and is the author of more than 20 patents on UWB technology.

Naiel Askar is the chief systems engineer for the advanced wireless group at General Atomics (GA). He has been an active participant in the UWB standardization efforts in the IEEE 802.15.3a and 802.15.4a working groups and the WiMedia Alliance. Within GA, he is the systems lead for UWB chip development. Before GA, he held lead systems design positions at a number of companies specializing in cellular and satellite communications and chip development. He received a PhD in electrical engineering from the University of Wales, United Kingdom, in 1984.

Jaiganesh Balakrishnan is a senior systems engineer with the Wireless Solutions Group at Texas Instruments in India. He received his PhD in 2002 from the School of Electrical and Computer Engineering at Cornell University in Ithaca, New York, and his B. Tech. in 1997 from the Indian Institute of Technology in Madras. Dr. Balakrishnan codeveloped the time-frequency interleaved OFDM (TFI-OFDM) UWB physical layer proposal for high-speed wireless communications in personal area networks. The TFI-OFDM proposal served as the basis for the WiMedia MB-OFDM physical layer specification. His research interests include detection and estimation, adaptive signal processing, and wireless communications.

Anuj Batra is a member of the Group Technical Staff at Texas Instruments, where he currently leads the IEEE 802.11n wireless LAN standardization and development activities within the DSP Solution Research and Development Center. He spent 2002 through 2006) leading the UWB standardization and development activities within Texas Instruments. During that time, Dr. Batra coinvented the MB-OFDM physical layer standard. He also served as the technical chair and editor of the physical layer task group within the MB-OFDM Alliance (MBOA) and WiMedia Alliance. He received a PhD from the Georgia Institute of Technology in 2000. Since then, he has been with Texas Instruments. From 1992 to 1993, he worked at Raytheon E-Systems in Falls Church, Virginia. In 2004, Dr. Batra was named one of the world's 100 Top Young Innovators by *Technology Review*, the Massachusetts Institute of Technology's magazine of innovation.

Jason Ellis is part of a small team responsible for making UWB a commercial reality. He joined Staccato Communications in its early stages working in business development and marketing; having previously led business development and marketing at General Atomics Advanced Wireless Group. He was instrumental in the founding and development of UWB standards, from the MultiBand Coalition, MultiBand OFDM Alliance, and now the WiMedia Alliance. Involvement in IEEE standards is extensive, from technical editor for IEEE802.15.3a, vice chairman and cofounder of IEEE802.15.4a, and the Working Group 802.15 Executive Committee. Further, he served on the board of directors for the 1394 Trade Association, is involved with the Consumer Electronics Association, Certified Wireless USB, and served as a regulatory delegate to the ITU. Previous experience includes tenure at Raytheon Systems, where he designed and implemented networks including WLAN and VoIP. He holds five UWB patents, a bachelor of science in electrical engineering from the University of California, San Diego, and has studied international strategic business at Oxford University.

Heinrich Foltz is associate professor and department chair of electrical engineering at the University of Texas, Pan American. He received his BSEE, MSE, and PhD degrees from the University of Texas, Austin. His research interests include reduced-size antenna design, broadband antennas, and RF circuits.

David Furuno is the director of the Advanced Wireless Group at General Atomics. Prior to founding the group in 2000, he held positions in the management of advanced programs at Lockheed Martin, Martin Marietta, and TRW. He received his PhD from the University of California and his SB from the Massachusetts Institute of Technology.

Susan Lin is principal systems architect in the Advanced Wireless Group at the Photonics Division of General Atomics. She received her PhD in 2000 from Stanford University. She specializes in applying digital signal processing and communications theory to telecommunications and magnetic recording systems. Prior to joining General Atomics, she was a communications engineer at Fantasma Networks, managing the PHY testing and integration of an UWB wireless system.

Michael McLaughlin is chief technical officer at Decawave and has worked for over 20 years in electronic communications. Before founding Decawave, he was the chief technologist at Cornel Electronics and LAKE Datacomms, and while there, he contributed to the V.34, V.90, and V.92 modem recommendations. He invented the technique used in V.92 known as compound precoding. Decawave, along with Freescale and the National Institute of Information and Communications Technology (NICT), are coauthors of DS-UWB, one of the two UWB proposals considered by IEEE802.15.3a. He is a member of the IEEE802.15.4a task group and proposed the preamble and convolutional code selected by the group for that UWB standard. He also helped to define the selected modulation scheme.

James S. McLean (IEEE member 1990, IEEE senior member 2004) received his BS, MS, and PhD degrees in electrical engineering from the University of Texas, Austin, in 1984, 1986, and 1990, respectively. He was a member of the technical staff at TRW Antenna Systems Laboratory in Redondo Beach, California. Since 1998, he has been with the TDK Corporation in Cedar Park, Texas, performing research in antennas, electromagnetic compatibility, and electromagnetic metrology. His technical interests include broadband, electrically small, and low-profile antennas, and he has carried out research in these areas in the Electrical and Computer Engineering departments at the University of Wisconsin, Madison, and the University of Texas, Pan American.

Charles Razzell received his undergraduate electronics engineering education at the University of Manchester Institute of Science and Technology in the United Kingdom in 1981. Subsequently, he has been involved in various wireless technology projects, usually involving integrated transceivers. His work for Philips has spanned two decades, initially in Europe and later in the United States. He has made regular technical contributions to the IEEE 802.15.3a task group and is an active member of several technical committees of the WiMedia Alliance. He has nine published U.S. patents and is a member of the IEEE Signal Processing and Communications Societies.

Robert Sutton is the president of TDK RF Solutions as well as an acting board member of TDK R&D Corporation. His undergraduate and graduate degrees are from the University of Texas, Austin. Before cofounding the business that ultimately became TDK RF Solutions, he worked in the electromagnetic compatibility laboratory at AT&T Bell Laboratories from 1988 to 1992. Much of his published work covers the topic of small radiating structures and electromagnetic compatibility and coexistence issues.

Larry Taylor is systems architect at Staccato Communications. He is one of the contributors to the Certified Wireless USB specification and is a member of the WiMedia MAC Specification Committee. He has served as committee chairman and technical editor on various standards for wireless data communications for more than 15 years and specifically on UWB communications over the last 4 years.

Index*

A
Active Transaction Period 261, 262, 265
Adjacent channel systems 252
Aiello 15, 71, 246, 313, 329
Alereon 9, 11, 291, 296, 328
Analog filtering 67, 171, 172
Anechoic
  chambers 19–22, 25, 26, 89, 90
  material 20, 22
ANSI 18, 28, 51
ANSI C63 20, 28, 29, 33, 51
Antenna
  designs 74, 113, 141, 329
  factors 27
  field factor 84
  gain 19, 41, 143
  impedances 84, 93
  input impedances 85
  length 122
  materials 96
  parameters 73
  response 102
  transfer functions 73, 85, 140
Antenna-to-Antenna Transfer Functions 83
Antipodal signals 148
Aperiodic cross-correlation 156
Archimedian spiral antennas 136

B
Assigned bands 236
Autocorrelation function 153–155, 157
Avoidance 54, 65, 66

Balakrishnan 15, 246, 247, 313
Balanced antennas 109, 113
Baluns 32, 109, 112–115, 141, 236
Bandwidth
  expansion factor 234
  measurements 25
  reduction 174
Barker sequences 157, 158
Batra 15, 246, 247, 313, 314
Beacon
  frames 265, 267–269, 275–277, 281, 282
  group 266, 267, 279, 281
Beacons 264–267, 273–278, 280, 281
Biconical antenna 105, 116, 121–128
Binary sequences 148, 155, 157, 159
Bluetooth 9, 12–14, 161, 290, 294, 297, 301, 304, 307, 328
Bluetooth SIG 289, 293, 296
Bowtie antennas 121, 122
BPSK 148, 151, 152, 160, 180, 181, 200, 207, 212
Broadband antennas 73, 314

---
* This index was created with **TExtract**™

C

Camera phones 301, 303, 306, 307, 310
Canonical antennas 74, 87
Capacity 2, 3, 147, 195–198, 250, 251, 277, 314
Captured multipath energy 217, 219
C-band 238
CDMA system 216, 217
CEPT 59, 60, 71
Certified WUSB 14, 294, 296
Channel
   capacity 161, 162, 196, 198
   estimation 225, 230, 231, 235
   impulse response 157, 165, 173, 217, 225
CISPR 18, 39, 42, 51
Coexistence 8, 50, 53–55, 57, 59, 61, 63–65, 67, 69, 71, 211, 239, 259, 296, 325
Common radio platform 294, 296
Common-mode structure 109, 114, 115
Communications
   channel 250, 252, 261, 267
   systems 5–7, 34, 147, 153, 249
Complex antenna factor 84, 142
Compliance 17, 28, 46, 163, 192, 294, 296, 325
Conical Antennas 91, 144, 145
Constant-aperture antenna 83
Constant-gain antennas 83, 91
Convolutional code 151, 152, 163, 166, 178, 227, 232, 315
Correlated energy 103, 105, 122–127
Cross-correlation 155, 156
Cylindrical Antenna 145

D

DAA 50, 54, 56, 61–66, 69, 70
DCM 226, 228, 229, 237
Despreading 175
Detect 5, 64, 66, 176, 231, 255, 269
Detection 6, 8, 54, 65, 66, 178, 288, 313, 325
Detector 29, 44, 108, 160, 204, 209

Digital
   baseband 212, 217, 224, 236, 237
   cameras 14, 299–301, 303, 304, 307, 308
   device 19, 36, 41, 43, 302
   filtering 172, 173
Dipole 104, 109, 110, 114, 116, 124, 128–130, 132–134, 137
   antenna 114, 137, 143
Direct sequence 147, 324, 325
Direct-Sequence UWB 11, 147, 149, 151, 153, 155, 157, 159, 161, 163, 165, 331
Dispersion 92, 99, 100, 105, 112, 114, 115, 136
Distortion 101, 104, 136, 161, 201
Diversity 208, 227, 228
DS-UWB 69, 147–149, 159, 160, 162, 292, 294–296, 298, 315
Duty cycle 65, 185–187, 310
DVD 222, 302, 303, 306, 310

E

Effective length 83–85, 97, 101
EIRP 19, 40, 41, 44, 45
Electrically Small Antennas 142, 144
Elevation Angle 106, 123–127, 140
Emission limits 8, 56, 238
Emissions 18, 30–33, 35–38, 41, 43, 48, 51, 122, 170, 222, 238, 242, 243
Energy
   efficiency 73, 92, 96, 98, 102, 128, 130, 132, 135, 300
   gain 73, 97, 98, 105, 106, 123–127, 133–135, 139, 140
   radiation efficiency 92, 94, 96
Equalization 100, 102, 107, 108
Equalizers 104, 165, 194, 208, 216, 217
Europe 49, 51, 53, 54, 58, 61–65, 69, 237, 245, 315, 328
Exposure limits 47, 48

F

Fano 76, 78, 142

FCC 2, 7, 8, 17, 18, 28, 29, 46–50, 56, 58–60, 70, 74, 153, 184–187, 211, 244, 245, 323, 324, 327, 328
FFT 5, 66, 67, 218–220, 222, 224, 225
Filters 27, 82, 117, 154, 163, 165, 171–173, 222, 236
Fingers 180, 181, 217
Finite bandwidth antenna 93
Fixed
   antennas 82, 85
   aperture antenna 37
Foerster 245, 246
Fractional bandwidth 18, 19
Freescale 8, 293, 294, 297, 315, 328
Frequencies of operations 2, 66
Frequency
   bands 18, 35, 53, 54, 65, 195, 241, 325
   domain 24–28, 69, 95, 96, 115, 143, 167, 168, 170, 203, 234, 239
   hopping 48, 49, 161
   range 20, 23, 29, 34–36, 38, 42, 43, 52, 54, 61, 73, 74, 108, 119, 122, 135, 138
Frequency-independent antennas 115, 135, 136

G

Gaussian pulse 106, 129, 130, 136
General Atomics 9, 10, 191, 209, 290–292, 313, 314
Gold sequences 158
Ground plane 20, 37, 88–91, 109–112, 117, 119, 127
Ground-plane independence 107, 109, 110, 112, 141

H

Handheld devices 19, 35, 224
Harmful interference 1, 8, 53, 54, 60, 324
Harmuth 3, 4, 6, 15
High data rates 194, 250, 251, 290
High-gain antennas 73, 81
Hub 255

Human interface devices 303

I

IEEE 2, 6, 9, 11, 15, 22, 51, 117, 141, 209–211, 244–246, 282, 285, 289–298, 313–315
Impedance bandwidth 74, 76, 105, 110, 117, 118, 121, 144
Impulse 69, 70, 157, 159, 168
   radios 1, 9, 11, 54, 66, 69, 70, 159, 192, 291, 292, 323–325
   response 3–5, 24, 157, 244
Indepen 55, 60, 70, 71
Indoor 32, 34, 40, 41, 44, 46, 48
Industry Standard 9, 285, 324, 326, 329
Infinite
   balun 113
   ground plane 75, 110, 111
Information theory 166, 195, 196, 210
Insertion loss 90, 91, 114
Instantaneous bandwidth 161, 162, 217
Intel 9, 10, 289–292, 294, 301
Intentional
   emissions 42, 44
   radiator 18, 19, 29, 30, 35, 36, 41, 43, 46, 47
Interference 2, 6, 10, 17, 44, 48, 49, 53–55, 61, 63–65, 69–71, 176, 177, 185, 187, 234, 324, 325
Interferers 176, 214, 235, 236
Interval Research 7, 8, 291, 313, 323, 329
IP 12, 13, 250, 294, 296, 299, 301, 308
ISI 153–155, 159, 192, 194, 208, 216, 217
Isotropic antenna 19, 41, 107, 108, 229
ITU 58, 60, 71

J

Japan 50, 51, 54, 61–66, 69, 214, 237, 238, 245, 247, 294, 328

L

Large bandwidths 2, 140, 167, 212, 300, 325

Linear
  dipole antennas 87
  reciprocal antennas 95
Link budget 24, 214, 229, 230
Loaded Antennas 128
Loaded dipole 131, 133–135
Local oscillators 165, 168, 174, 175, 177, 179, 188
Log-Periodic Antenna 145
Low-profile antennas 315

M

MAC 12, 173, 222, 231, 246, 250–254, 260, 271, 281, 282, 289, 294–296
Market 13, 31, 51, 53, 54, 214, 285–287, 290, 293, 297–301, 305, 307
  requirements 286, 287, 298
Mass storage 304, 310
Matched filter 156, 159, 165, 166, 173, 204, 205, 208
MBOA 10, 11, 245, 292, 294–296, 314
MB-OFDM 54, 212, 213, 220, 222, 224, 234, 244, 245, 313, 328
  system 212, 216–235, 237, 239, 240, 242
MBOK 150–152, 154
McLean 52, 73, 118, 142–145, 315, 331
Measurements 19–22, 24–30, 33–38, 40, 42–44, 46–51, 57, 75, 88, 142, 184, 189
Minimum-distance detection 204, 205
Mobile phones 11–14, 144, 293, 298, 304
Modulation scheme 91, 92, 148, 161, 162, 215, 218, 325
Monopole 75, 84, 88–91, 108–111, 116, 118, 121
MP3 players 301, 302
Multiband 9, 10, 170, 174, 192, 212, 291, 292, 325
Multiband OFDM 211, 213, 215, 217, 219, 221, 223, 225, 227, 229, 231, 233, 235, 237, 245–247
Multiband systems 175, 184, 187, 188, 211, 325
Multipath 10, 83, 92, 166, 208, 220, 225, 226, 235
  channels 166, 208, 216, 226
  energy collection 178, 188, 213
Multiple sub-bands 167, 191, 192, 209

N

Narrowband 24, 76, 78
  antennas 74, 98
Near-field coupling 110, 115
Networks 3, 74, 78, 82, 114, 115, 266, 299, 301
Noise 24, 82, 100, 160, 161, 196, 201, 229, 252
  bandwidth 160, 161
  figure 214, 229, 230
Noncoherent receiver 208, 209
Notch 62, 64, 67, 69, 176, 238–242, 244
NTIA 49, 56, 57, 188, 189, 324, 327, 328

O

OFDM 9, 10, 70, 92, 162, 169, 170, 181, 213, 216, 220, 245, 291, 292, 324, 325
  system 10, 218, 219, 224, 225
Omnidirectional antennas 92, 142
Operating
  bandwidth 213–216, 222
  frequencies 110, 312
Orthogonal sequences 149
Oscillators 182, 183, 223

P

Parallel receivers 179, 180
Patent 1–3, 5–7, 313, 315
PCA 268, 279, 280
Peak-to-average ratio 153, 213, 226
Peripherals 13, 14, 32, 33, 299, 302, 304, 305
Personal
  computing 12, 245, 290–293, 298
  devices 299, 302
Petition 48, 52, 184

Phase 99, 100, 107, 136, 137, 182, 192, 193, 199, 200, 205, 207, 212, 263
PHY 9, 11, 12, 231, 236, 250, 252–254, 260, 263, 268, 272, 281, 282, 289, 294, 296, 314
Physical layer 9, 11, 212, 216, 250, 282, 289, 314
Piconets 168, 221, 233–236
Pilot tones 222, 226, 244
Portable
  devices 11, 19, 301, 303
  media players 300, 301, 307
Power
  consumption 2, 161, 167, 174, 175, 208, 212, 213, 236, 237, 293, 300
  levels 61, 62, 227
  spectral densities 2, 50, 51, 59, 170
  spectrum 170, 172
Printers 14, 299, 300, 302–304, 309, 310, 329
Processing gain 160–162, 217, 226, 234
Proprietary 287, 288, 290, 293
Pulse
  distortion 99, 100, 115, 130, 141
  energy 93, 95
  length 133, 134
  shape 78, 88, 92, 98, 104, 122, 123, 135, 154, 161, 170, 172
Pulsed
  antennas 112
  multiband 169, 212
Pulse-position modulation 7, 212

Q
QPSK 10, 180, 193, 220, 226, 228, 237
Quality factor 28, 142
Quotient 55, 60, 70, 71

R
Radar 4, 6, 15, 73, 140, 159, 324
Radiated
  emission limits 30, 31
  emissions 30, 34, 42–44
  measurements 29, 35, 36, 43, 50
Radiated-emissions measurements 21, 22, 29, 35
Radio 2, 14, 24, 31–34, 37, 44, 48–50, 54, 58, 66, 75, 82, 109, 115, 163
  astronomy bands 238, 239, 241, 242
RAKE
  fingers 179–181, 212, 213, 216–218
  receiver 208, 216–218
Reactance 111, 112
Regulations 2, 17–19, 21, 23, 25, 27–33, 35, 37, 39, 41, 43, 45–47, 49–51, 65, 66, 237, 238
Regulators 31, 56, 63, 65, 184, 329
Reservation 264, 268, 269, 277–279
Resistive loading 77, 78, 96, 128, 132
Resolution bandwidth 24, 29, 40, 42, 44–46, 242
Ring oscillators 182
Ross 3, 5, 6, 15
Rules 17, 18, 32, 34, 45–47, 49, 184, 196, 259, 279, 324

S
Sampling rate 99, 170, 173, 175, 219
S-band 238
Security 22, 260, 261, 295
Shift register 157
Short range 253, 299, 300, 324
Short Range Devices 52
Signal
  bandwidth 161, 208
  power 160, 161
Silicon 11, 182, 292–295, 299
Single-carrier system 216, 218, 219
Small antennas 74, 76, 108, 129, 140, 142
SNR 148, 160–162, 250
Source impedances 93, 96, 97, 105, 129, 130
Specifications 2, 9, 11, 51, 244, 249, 254, 263, 282, 283, 285, 287, 288, 293, 294, 296, 297, 315

Spectral flexibility 213, 219, 237, 238
Spectral Keying™ 191, 193, 195, 199, 201, 203, 205, 207, 209, 331
Spectral mask 61
Spectrum 1, 2, 8–10, 36, 37, 53, 56, 57, 63–66, 69, 147, 154, 160, 161, 170, 211, 212, 238, 239, 241, 242, 323–325
  allocations 8, 61, 64, 237
  analyzer 23, 27, 30, 69
Staccato Communications 8–10, 12, 290, 292, 294, 296, 313–315, 329
Standards 1, 32, 50, 249, 285–287, 289, 291, 293, 295, 297, 315, 324, 326, 331
Sub-bands 167, 169, 173–178, 181, 182, 184–186, 191–194, 202, 203, 208, 220, 222, 223, 239
Superframe 265, 266, 268, 270, 271, 273, 275–277, 281
Supergain antennas 81
Switch 55, 163, 222, 223, 272, 277
Symbol rate 160, 161
Symbols 100, 148, 150, 153, 161, 162, 177, 192–196, 198, 199, 201–204, 207, 208, 211–213, 228, 253, 272, 281

T

Tapered dipole 74, 75, 108, 110
Task group 11, 15, 244, 246, 287, 315
Ternary codes 148, 158, 159
Texas Instruments 9, 10, 313, 314
Time domain 1, 4, 5, 7–10, 25, 69, 95, 96, 102, 115, 143, 153, 208, 225, 234, 289, 323
Time-domain OFDM symbols 220, 221
Time-domain pulses 99, 100, 136
Time-frequency codes 221, 222, 234–236
Tone nulling 67–69, 239, 240
Tones 67, 219, 224, 226, 228, 229, 239, 241, 242
Transfer function 73, 83–88, 91, 99, 100, 136

Transmit
  antenna 85, 88, 93, 96, 101
  power 2, 62, 64, 66, 149, 150, 163, 166, 173, 184, 212, 213, 226, 238, 242, 325
  spectrum 10, 62, 67, 154, 215
Transmitted OFDM symbols 240, 244
Trellis diagram 151, 204–206
Triangular antennas 90, 144
TV 14, 303–305, 307–311

U

Unlicensed communication devices 56
USB 12–14, 32, 253–255, 266, 290, 295, 304, 305, 307, 309, 311
USB-IF 289, 293, 295
UWB
  antennas 73, 76, 78, 80, 81, 83, 91, 106, 107, 112, 115, 121, 140
  bandwidth 18, 32, 37, 49, 53
  devices 17, 28, 29, 31–36, 38, 46–48, 54–59, 61, 64, 65, 211, 219, 238, 251
UWB Forum 289, 293–296
UWB
  radios 11, 17, 28, 32–34, 66, 67
  spectrum 1, 2, 7–9, 176, 191, 192, 222, 238, 313, 324
  transmitters 3, 7, 17, 18, 34, 41, 42, 47, 49, 54, 64, 68, 325

V

Valentine's Day 327, 328
Video 1, 14, 55, 280, 299, 302–304, 307, 308
  bandwidths 27, 40, 42, 44

W

Waiver 47–49, 52, 184, 328
Walsh codes 156, 157
Wearable devices 323
Wheeler 76, 110, 142
WiFi 12, 14, 304, 307

WiMAX 53–55, 61
WiMedia 11, 245, 253, 265, 282, 294, 296, 298
WiMedia Alliance 9, 11, 245, 282, 289, 293–297, 313–315
WiMedia MAC 250, 266, 267, 271, 273, 279, 282
Wired USB 14, 254, 255, 257
Wireless systems 67, 162, 163, 324, 326
Wisair 9, 10, 13, 289, 291, 292, 294
WPANs 249, 282, 289, 299, 302, 303
WUSB 11, 12, 14, 34, 249, 253, 254, 257, 259, 260, 263, 265, 266, 279, 290, 299, 301, 305, 307–309

X

XtremeSpectrum 8, 289, 290, 292–294, 323